DAYS OF STEAM AND GLORY

by
Dana
Close
Jennings

NORTH PLAINS PRESS
ABERDEEN, S. DAK.

Minneapolis engine on bonanza farm pulls 14 plows.--From "The Day of the Bonanza," by H. M. Drache, courtesy the author and his publisher, N. Dak. Institute for regional studies.

Contents

Dedication

This book is dedicated to the hard-working folks of Madison and Lake County, South Dakota, whose hardy forefathers tamed the wild prairie and made it fruitful——banishing the spectre of famine for the first time in human experience——and whose own dedication, courage and sweat are re-creating Prairie Village as a living memorial to their sires.

Madison's Prairie Historical Club held its first steam threshing jamboree in 1963 with this one steam engine borrowed from Adolph Rude, Flandreau. Club members now own six steamers. "I could have bought it three years earlier for $500," Rude ruefully told me in 1963. "But I thought the price was too high. So I waited for the price to go down. The next year it was a thousand bucks. I finally got it for $1,500, the price for a new one." The 40-HP Case was 60 years old when he bought it.

Introduction

If your nerves have not tingled to the smell of steam and hot cylinder oil--if your blood has not raced with the rhythmic chuff-chuff of a steam traction engine under load--you have not fully lived. But cheers!

You yet may live.

These nearly-forgotten sensual delights can now be yours at any of the many steam threshing jamborees up and down our land where old settlers meet, old threshermen thresh, and the shout still echoes across the land:

"Hold Your Horses!"

...and the steam whistle splits the dewy dawn.

Each summer, even in these space-age times of ours, old-time steam threshing bees attract increasing thousands of spectators. Old-timers thrill again to throbbing steel under their feet--to the bite and rustle of the grain as they cut the bundle-bands and hand-feed an ancient threshing machine. Youngsters thrill to purposeful uproar, seeing how their grandfathers worked and lived. Grandmothers brag to their daughters-in-law of the great threshing-day feast they prepared over the roaring, smoking, cob-burning range in the stifling oil-lit kitchen.

A century ago our forefathers ate by the sweat of mule and man and horse. Then steam left the rails and marched across the fields; the first relief human and bestial muscles had gotten since man first poked seed into stubborn soil. Steam sparked and powered the Agricultural Revolution for more than half a century until the clanking, stinking gas engine displaced it.

Steam was King from soon after the Civil War to the First World War, when food demands stimulated the lighter, faster, more nimble gas tractor. During the glory of steam, the Engine Man was King of the farm community, at least during the four to six weeks' threshing. Mothers fearfully warned their daughters against the blandishments of threshermen, whose very prestigiousness made them all the more alluring to farmers' daughters.

Steam and gas battled for another decade, when weight and honored age gave way to speed the noisome youth.

Farm steam has now traversed its obsolescence and is well into its antiquity. Steam hobbyists are today rescuing mountains of rust from swamps and junkyards and putting them into mint condition. Collecting, restoring and running these gentle giants——some topping 30 tons and 60 years——is the fastest-growing farmers' hobby from Pennsylvania to the Pacific.

You, too, can feel your heart leap to power and neighborliness, sniffing the inimitable steam-smoke-oil perfume, itching to the chaff down your sweaty neck, hearing the old sounds and tasting the dust of days long gone: just go to a Steam Show.

This book is the distillation of the knowledge, skill and experience of many people. Thanks go to Palmer Reinicke, farmer, Winfred, S. D., who provided and checked the engine specifications and other technical matters; to George Klinkner, farmer, Artesian, S. D., who with his son Don braved the icy winds of early March to fire up their Case steam traction engine especially for the photos accompanying the steaming-up section; and for Mr. Klinkner's technical help on that section, thanks; to Prof. Joe Habeger, General Beadle State College, Madison, S. D., and president of the Prairie Historical Club, whose encouragement and technical help were invaluable; and to all the members and earnest workers of the Prairie Historical Club who built Prairie Village and made this book possible.

Particular thanks are due to all the operators who"held that Pose" and "did it again" for pictures, and especially to the engineers who, no matter how it hurt them, "made 'er smoke" for photos.

Real steam buffs will travel half across the continent to enjoy a Steam Show somebody says is good, or just to see a rare engine. Some take pictures and some make recordings of the steam engine's powerful sound.

These Steam Days are the biggest days of the year in many a farm-country county seat, pushing the county fair into oblivion.

Clap on your straw hat and join the Fun Crowd!

We'll take you back to the Glory that was Steam.

This is a more frequent scene these days as old steam threshermen——and new ones, too——all across the land resurrect a nearly-forgotten art. This and most of the photos in this book were taken by the author at Prairie Village, Madison, S. D.

"The threshers are comin'! "

We kids use to lie in the weeds along the road ditch, wiggling our bare toes in the dirt, watching the steam engine come down the road."

Thus recalls Vern Laustsen, Aberdeen, S. D., publisher of THE DAKOTA FARMER and of this and other such notable books as CATTLE ON A THOUSAND HILLS by the same author. Born on a Dakota farm, and though even now his eldest children are only in college, he remembers Thrashing Day well. (Those of Teutonic stock on the Northern Great Plains will remember it as Trashing Day.)

The huge engine, belching smoke to the skies, pulling the threshing machine and maybe a water wagon and coal tender behind that, made a train immense to little eyes whose horizon had been the dirt road before them.

To the engineer, up at three to stoke the fire and get up 125 to 155 pounds of steam, it was a point of pride to reach the next farm in the threshing ring, get the separator leveled, the engine belted up and ready to go before the hot July or August sun burned the dew off the bundles. The kids heard the whistle rolling over the hills with the crack of dawn and ran to watch and wait among the dewy cucklebers.

Closer, closer the gigantic apparition lumbered at top speed––2.61 mph––shaking the very earth.

"Finally," Vern smiles, "the excitement got too big. One of us bolted through the horseweeds higher than our heads yelling, 'Ma! Th' thrashers 'r comin'!'"

As if Ma hadn't been up since three herself, peeling potatoes and picking chickens and baking bread and pies to feed the crew, which might total a dozen or two-score ravenous men, depending on the size of the separator to be "kept up in the collar," as we said of hard-working horse, machine or woman.

Dad was up even before Mom, routing out the boys to do the chores, to fetch the mountains of kindling, cobs and water Mom needed to feed the big black cast-iron range where pots of molasses-rich beans, choked with chunks of fat smoked pork and tangy with mustard, catchup and vinegar, had been simmering for three days "to make 'em good."

"That was the most exciting day of the year, bigger even than Christmas," smiles Vern fondly. "More bustle and noise and action than the little kids could remember. We big kids remembered the last Thrashing Day or two and we knew it was a Fun Day––fun for kids, anyway."

Men still remember Thrashing Day as the greatest day of the year, but the women remember mostly the day of the big work, the big sweat and the big sink full of dirty dishes.

"Thrashin' Day made Christmas Day and all the other days possible. It was the threshing machine that put the grain in the bin that put the money in the bank that filled the Christmas stockings and gave another year's grace on the mortgage," so recalls Vern Laustsen.

As the monster approaches, neighbors converge from all directions, some in wagons, some in buggies, some in rattling/gasping Model Ts, for on Thrashing Day everybody helps everybody. Neighbors "trade work" without quibbling about "I only had 20 acres for your 50." Communities formed threshing rings: one farmer with more capital or credit than the rest would invest $2,000 to $2,500 in a steam engine and separator and go from farm to farm to thresh the wheat, the oats, the barley and rye––rarely flax and clover––for 3 cents to 6 cents a bushel. (An equivalent combine today costs $10,000, farmland costs five times as much and grain prices are about the same today as they were then, as are threshing-combining fees.)

The excitement of a steam engine is more than the smallfry can contain. This little fellow streaks around trying to take it all in at once. You'll feel the same way if you ever get bit by the steam bug.

Neighbors help each other pitch, haul and scoop on Thrashing Day, no favors asked, no records kept. It's your way of life.

The men drive teams and flat-bed bundle wagons with a three-tined bundle fork wedged into the rear standard to pitch the bundles from the shock ("stook" farther north and in Canada) to the wagon, from the wagon into the separator. Those lucky farmers with a grain-tight wagon box are asked to haul the grain to the barn and scoop it through high, tiny granary windows while the hens willingly salvage their misses.

Stout wives bring covered dishes of their respective specialities—"tater" salad or graham cracker pie or pickled pigs' feet—and help Ma cook and serve and "red the table" (that's Wisconsin for "Rid the table of dirty dishes"). Were the house a big one, we've put all the leaves into the big round oaken dining table, set up extra tables and chairs in the parlor and on the porch and (as a last resort) in the Sunday parlor.

When the house can't hold the whole crew, we set planks on sawhorses for tables and benches out under the trees. And the woman who offers paper plates or store-bought bread is snickered at forever.

Feeding the threshers is a point of competitive pride among the women, each striving to set the best table. Her man is sure to have more help than he really needs. The farmer whose wife isn't much of a cook has a time rounding up enough reluctant helpers to keep the big machine up in the collar. And the woman who expects neighbors to bring sacked sandwiches—her man might as well sell out and move the lazy wench to town.

The cast iron pump, just outside the kitchen door, is a busy, noisy place as the men troop in from the field at noon, dust and sweat mortaring chaff to their faces, arms and across the wet backs of their faded blue shirts. More planks set up in the shade carry tin basins or tubs and soap and worn-out towels, saved just for Thrashin' Day to keep axle grease off the good ones. Scorning hot water to cut the chill of the well, the thrashers "dipper" fresh well water out of buckets, sloshing it on their faces, over their necks and behind their ears, rubbing hard, blowing and sputtering through the suds and making a great shower on the grass (a good reason, in the womenfolks' view, for setting the wash-up outdoors), rasping horny hands together to get the worst of the grease off. After all, a fella don't have to get too clean, just to eat, tain't like goin' to church.

The warm smells tantalizing dusty noses are myriad: the aroma of roasting meat and baking pies, the yeasty aura of bread hot in the oven plus the sharp vinegar tang of any pickles, relishes or other spicy stuff the cook might have taken a notion to fix at the last moment.

Just as fine perfume is based on odoriferous musk, so the fragrances of Thrashing Day Dinner play against a background bouquet common to all farmhouses of the day: an exhalation of kerosene for the lamps, soaked into the floors and boiled out by the heat. Second only to the coal oil is the aroma of sour milk which, again, emanates from the woodwork, mingling with wood smoke, some wisps of which escape the cookstove every time someone lifts a lid to poke in another stick. Then there's the essence of honest sweat and the barnyard scents tracked in on boots, plus the perfumes of thousands of feasty family meals.

Such a pungent mix, spelled out on paper, lacks the heart-string yank of the real thing—but to all noses that ever lived it, it was incense bespeaking shelter from the storms of weather and of life, where there was always hot food for the body and warm nutriment for the soul.

These old-time farmhouse smells compounded the cozy balm of home.

Threshing took, besides the engine crew, a separator man, bundle haul[ers] and pitchers, grain, coal and water haulers. Field is now a residential a[rea]

Spikers pitched bundles from shock (sometimes called stook) to 2-[...] wagon. This farmer is cheating—he replaced his old iron-tired woo[den] wagon wheels with auto wheels.

The Thrasherman's Dinner was always the same yet different: huge beef-roasts done brown plumb to the bone; platters of fried chickens, crisp and golden, that had trumpeted up this very dawn; snowy mashed potato mountains cratering lakes of home-churned butter without any artificial color in it, and more bowls of the same golden stuff heaping around; boats of rich brown beef and chicken-giblet gravy; platters of home-baked bread, still hot, and of steaming home-smoked ham; great brown earthenware crocks of much-baked beans lumy with real pork.

Then there were all the trimmin's: home-made pickles—watermelon, dill, mustard, onion, tomato; home-made fruit jams and jellies of wild plums, cherries, gooseberries, elderberries, grapes, whatever the woods and fields afforded. Fresh green onions and "reddishes" and leaf lettuce pulled from the garden by the door while the dew was still on them, along with tomatoes picked this very morn and kept cool in wooden tubs of chill well water or lowered into the springhouse, were the farm lucky enough to have such; deviled eggs picked yesterday and still warm from the stove; huge blue granite pots of brawny coffee; cream thick enough to cut.

And it's a lazy woman who doesn't offer for dessert apple, raisin, pumpkin, peach, cherry and mince pie, all hot. If she has a bit of humor in her, she'll put out at least one vinegar pie.

"Everything on the table come from our own place," invariably brags Dad, "'cept th' coffee, sugar, salt 'n pepper—'n Ma's eggs bought them!"

Dad, host, wastes no foolishness on a roast-carving ceremony. Hungry thrashers can't sit still for frills. The wise cook plops the roasts on the table torn to chunks ready to pile into, swimming in their juices in the big platter her granma brought over from the Old Country. Dad's job is mostly to keep things running—to show the separator man where he wants the strawpile, to show the grain haulers which bins to scoop into and to make sure there's hay and oats and water for all the neighbors' horses. He takes his turn pitching bundles, just to prove he isn't trying to get out of work, then scurries about other errands relative to supplying the engine with 2,000 or so gallons of soft water it will puff away today, plus the 1,500 or more pounds of coal.

Occasionally he stops long enough to peer up at the lordly separatorman, high atop his mechanical kingdom, sceptered with a long-spouted pump oil can, to ask, "How's she turnin' out, Emil?" and to grin at Emil's estimate of "a good thirty bushel."

One of the few remaining steam rings in the Red River Valley is on the S. A. Roen farm, 2½ miles from Comstock, Minn., along the Red River. Photo taken August, 1946.

"Dig in, boys. We ain't elegant," says Dad.

"Root, hog, or die," someone says ritually, licking his fork and spearing another drumstick.

"Fingers wuz made before forks," another inevitably says, tearing at a gold-dripping second joint.

Then there's not much sound for the first round except the clank of eatin' tools on thick china, earnest chomping, slurping tasty fingers and the murmured "pass the biscuits, Sam," and "say, Jake, chase some chicken up to this end, will you?" and "Oscar, slip the brakes on that-there roast beef."

After appetite's sharp edge dulls against seconds, the pace relaxes and conversation picks up.

"Feelin' better ev'ry minnit!"

"Man, is it hot!"

"You don't need a fire (this to the engineer) t'day. Jest open th' firebox door and let the wind blow through them tubes."

"By gum, it's so dry we hafta soak th' hogs over night before they'll hold swill."

"You think it's dry here? Got a letter frum m' sister in Kansas 'n th' stamp wuz **pinned** on!"

"You can count on a steam enjine. Yon Yonsen couldn't even git his gas tractor started. His missus is feedin' th' hull durn crew 'n not a bundle thrashed yit."

Someone is sure to recall the hapless neighbor's unfortunate wife in a Catholic community who, expecting threshers on Thursday, stocked up on beef roasts. That new-fangled gas tractor contraption stripped its bull gear, delayed the crew a day. She had to throw out all that good beef and buy fish.

After seconds, the women, damp from the cookstove's blast (belying the prissy dictum that a horse sweats, a man perspires but a woman only glows), pressed thirds upon the men: "Here, Hiram, have some more taters 'n giblet gravy. More chicken? Lan' sakes, Ole, you et like a bird."

"Ya, yoost a peck at a time," agrees Ole.

"Et so much makes him poor t' carry it," volunteers his best friend.

"I've got another pie in the kitchen," says Ma. "I'll cut it if anybody wants some."

This is a risky maneuver, especially if she's so unfortunate as to say it backwards: "I won't cut it if nobody don't want none." She thereby leaves herself wide open to behind-the-barn accusations of not wanting to put out more pie, and the tale will inevitably be told of the woman who, by this dodge, made one pie last all summer.

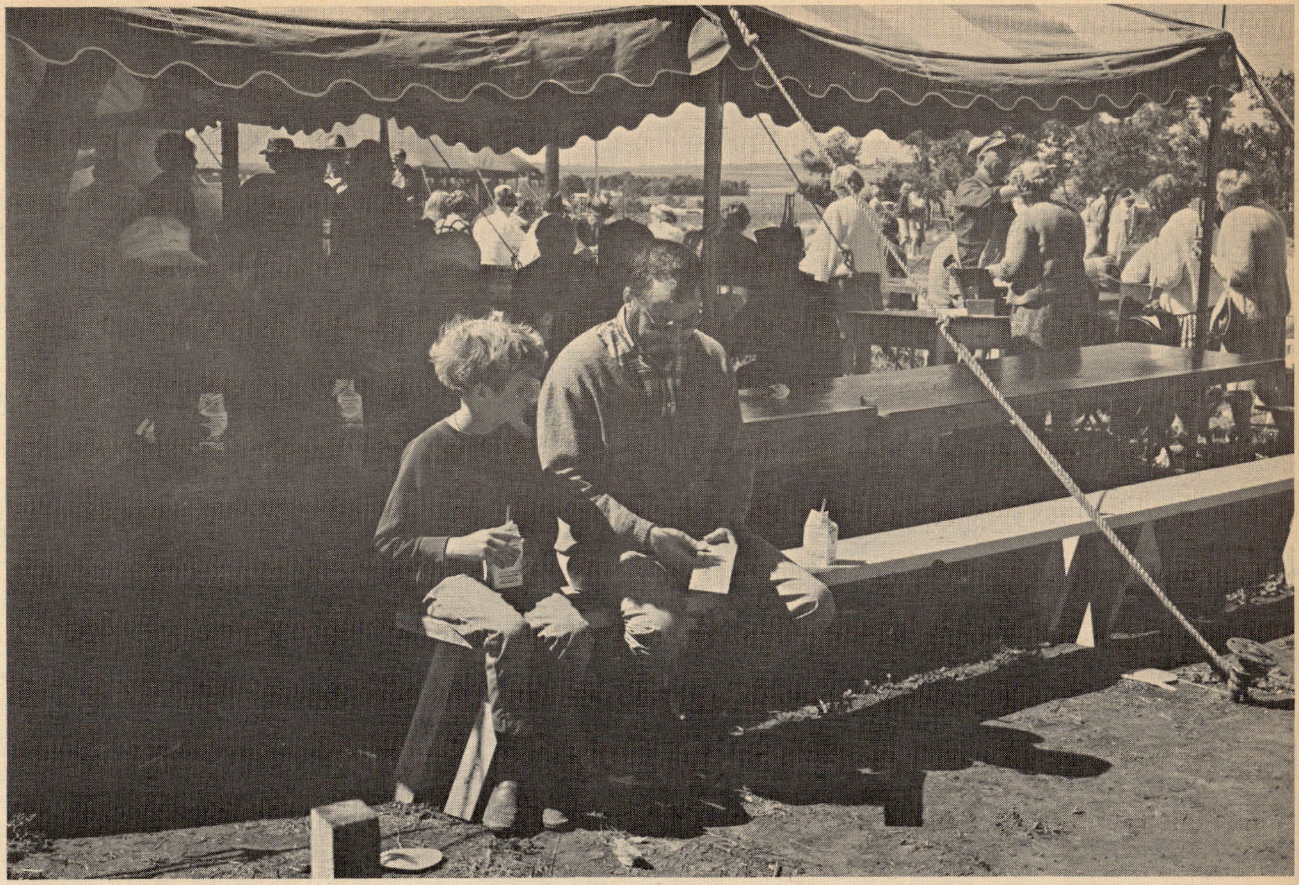

The sandwich stand you find at threshing shows falls far short of Mom's old-time thresher dinners, the Old-Timers assure you.

return-flue engine, web stacker and Gay 's-type bike mark this photo as a real oldy.

Old steel-wheel Hart-Parr tractor denotes early replacement of teams on bundle wagons ("Tractors ain't burnin' up hay while they're settin' still!") Soon after they replaced the steamer on the belt.

As the sun sank, bundles began to soak up atmospheric moisture and get too tough to thresh. Farmers had to quit and do chores anyway.

Picture is unidentified.

Boys big enough work with the men——scampering off to fetch a wrench, fill the burlap-soaked water jug, minding the water or the horses or the fire, breaking coal, even driving teams and pitching bundles. By virtue of this apprenticeship they get to eat with the men. The women and girls eat after the men but their pain is not too extreme; they've been tasting and snitching all morning. The vast reservoir of human misery lies in the tummies of the smallfry, who in this era before babysitters are ordered only to "keep out from under foot, now," and must wait for second table.

Following dinner there's a brief pause for rest, neighborin' and slurping toothpicks. (On the farm, even today, you eat dinner at noon. Supper comes in the cool of early dark, after chores. Lunch is that little snack of meat sandwiches, pie and coffee the women fetch to the field in mid-morning and mid-afternoon to pry your belt buckle loose from your backbone and to take the rubber out of your knees [lunch is not seen so much any more, now that the farmer is reduced to manager-machine operator]).

It took two wagons and four men pitching together to keep a big steam outfit up "in the collar."

As one child of these sweaty days recalls, "My folks raised and educated six kids on 160 acres, but I was 22½ years old before I knew Grace ended with just 'Amen' instead of with 'Amen-hitch-up-boys!'!"

One such meal a year was the average farm wife's obligation, besides helping all those neighbor women who helped her, for the average farm was threshed in a day, barring breakdown. Such a crew could clean up two little farms in a can-see-to can't-see day. It was the rare farm that had enough grain to keep the separator busy a second day. Of course, the big western wheat ranches owned their own equipment, hired their own crews.

Come choretime, the neighbors rattle home to milk their own cows by lanternlight, since you can't thrash after the straw toughens with evening moisture anyway. The engineman, who must be belted up again in the morning on another farm as soon as the dew is off, rolls in a blanket under the water wagon, sung to sleep by the engine's soft sighing to itself as little wisps of smoke and fugitive steam escape into the warm night. If he has some crew of his own, they sleep in the bright golden strawpile or in the hay loft over bins full of yellow grain promising another year's grace against "The Mor'gi-e." The pasture is full of tired horses. The beds are full of exhausted men, women and children. The hogpen is full of fat swine full of the most gorgeous heap of garbage within living pig memory.

Lining up the belt separated the men from the boys when it came to running a steam engine. You couldn't have the wind blowing chaff back on you so you tried to set windstacker downwind. When wind switched, you had to change the set to another direction. When you had to thresh in a strong crosswind, the engine might have to be two feet out of line to keep the 150-foot belt on. The belt looks dangerous, but I never heard of anybody tangling with it.

"We buy junk. We sell antiques"

That's a sign in the Kissin' Cousin Candy store between Madison, S. D., and Prairie Village.

Just when does a hunk of obsolete junk become a valuable classic?

Apparently, with steam engines, this subtle but important psychological change occurred during WW II. Junked engines became valuable. In the '40's old-time threshermen began to rescue them from swamps and junkpiles and lovingly to restore them. Perhaps so many got broken up during wartime scrap drives, people were shocked into realizing here was a precious bit of our recent past that would soon be irrevocably lost if Somebody Didn't Do Something.

Besides that, once a steam man, always a steam man. There's another subtle difference between playing around with an old-fashioned, obsolete machine—and keeping alive an ancient and noble skill.

Thus steam threshing was reborn as individual farmers, just for nostalgic kicks, fired up their old engines, invited the neighbors to help, and were amazed at the crowd of city people and reporters attracted by the smoke and din. Good old Yankee enterprise suggested charging a dollar admission.

The National Threshers Association, Inc.

While steam's rebirth has not been chronicled, apparently the first steam show was staged June 30, 1945, when LeRoy Blaker invited some steam enthusiasts to his farm near Alvordton, Ohio, to watch his three engines work. Three years later the National Threshers Assn. organized under Pres. Blaker. The 75 original members were from Ohio and surrounding states plus Utah. Since then, over 12,000 names have been recorded. Mrs. Earnest Hoffer, Toledo, Ohio, presides over the National Thresherwomen. Another subsidiary is the Baker Fan 500 Club.

NTA says it is the first association to use (1945) a Prony brake (horsepower testing device, a sort of a crude dynamometer) and first Baker fan (1954) at a threshers' reunion. They draw between 20,000 and 30,000 spectators to their four-day annual June show at Wauseon, Ohio.

Port Huron engines owned by NTA and LeRoy Blaker: 1921 19-65 HP No. 8635; 1917 24-75 HP No. 7948; 1920 24-75 HP No. 8503, winner three times in the economy run using less than 23 lbs. water per HP-hr.—Photo courtesy NTA

Case 60-HP 1911 with early 1900 tiller. Arthur Lucas, Clermont, Ind., owner.--Photo supplied by John J. Menchhofer, Indianapolis, Ind.

The Pioneer Engineer Club of Indiana

This was organized in 1948 with five men and one steam engine, a 50-HP Case. "Now our membership amounts to about 500 paid-up members from every state in the union including Hawaii and Alaska and Canada, France and England," reports John J. Menchhofer, sec.-treas., Indianapolis, Ind.

Today, he says, "We have about 25 large engines such as the New Huber, Case, Advance Rumely, Keck Gonnerman, Russell, Peerless, Minneapolis, Baker, Nichols & Sheppard. We have an 1876 wooden hand-fed Farquar separator, hand-sacked and web-stacked." Members own such gems as a 1900 wooden Case 36-inch separator, all kinds and makes of stationary gasoline engines, all kinds of model engines, separators and sawmills.

"We exhibit an old steam peanut and popcorn machine and an old hurdy-gurdy hand organ. Every year we have a 3-day show in August."

5-HP model Case engine operated matching sawmill at the Pioneer Engineers show.--Menchhofer

Model Case cranks out 5 HP, big Case, 50. The 1924 steamer is owned by Dr. Russell Holmes, Louisville, Ky.--Menchhofer.

John J. Menchhofer, secretary-treasurer, Pioneer Engineers Club, Indianapolis, poses by a 1925 18-HP New Huber return flue owned by Roy Tempest, North Vernon, Ind.--Menchhofer

Crowds at the Old Time show are attracted to 40-HP Case owned by Eiffel Plaster, Huntington, Ind.––Photo courtesy Jim Whitbey

The steamer with the fringe on top is a 1913 M. Rumely 20-HP, 10x10¼, 12 tons, Gould balance valve. Cost new about $2,000.––Whitbey

"They're a different-looking engine without the canopy," enginemen tell you, and you have to agree. Ralph Lindsay, Los Angeles, owns this 50-HP Case pulling a 36-60 Minneapolis separator in the Old Time parade.––Leo Clark Photo

1915 18-HP Advance Rumely owned by Jim Whitbey pulling 8-bottom plow.––Photo by Earnest Hoffer, Toledo, Ohio

Unidentified

The Old Time Threshers and Saw Mill Operators Ass'n. Inc.

This Indiana group started in 1951, reports Mrs. J. H. Whitbey, Ft. Wayne, Ind., when her husband, a retired railroad engineer, bought a 1913 Rumely 20-HP steam traction engine and a Baker threshing machine and put them into running order. He and F. W. Bloom, Churubusco, Ind., who had a 5-HP working model steam engine and threshing machine he'd made, put on a demonstration in Whitbey's wheat field along the road.

Jay Gould, noted farm director, WOWO, Ft. Wayne, announced it and, much to the Whitbey-Bloom surprise, produced a throng.

The third show brought 12 big engines, some small ones, four threshing machines, sawmill, Prony brake. "The show has grown now to include everything––not only 20 large steam engines and models, large and small, but Baker fans, gas engines, tractors, antique cars and farm machinery, steam plowing, veneer machine, teeter-totter, incline, threshing, corn husking and corn shelling," reports Mrs. Whitbey.

[On the teeter-totter, a steam engineer climbs his 12-ton monster onto a gigantic see-saw and, with a skilled hand on the reversing lever, balances her there. Sorry we couldn't find a photo for you.]

There's a show wherever you go

Up and down and across America you'll find these steam shows, threshing bees, etc., going on just about every day from May to October with an astonishing array of ancient equipment, all of it working like new. You'll find clubs from New York to California, rescuing and rejuvenating not only steam engines but early gas tractors, barrel engines, cars, farm implements, airplanes and early-day home and farm gadgetry.

You could leave home in May and cross the country clear into Fall and hit a steam show somewhere in every area. In many farm counties the annual steam threshing bee outdraws the county fair and becomes the biggest event––in terms of participants, spectators and tonnage of equipment––of the year.

Big daily feature of every show you find will be the parade. All machines have been restored to working order, many repainted in original reds, greens and gold.

Prairie Village, South Dakota

where the past comes alive

Prof. Joe Habeger, left, hands over the "key" of his 1915 Case 20-40 gas engine to Jerry Prostrollo, Prairie Village fund drive chairman. Joe's gift was the first property it owned.

One of the most remarkable——notable because of the tremendous strides it has made in its few years of existence, for the fact that an entire historic village has grown up around it, that it is a continuing operation, open all day every day from Memorial Day through Labor Day, and that it is designed as an educational-recreational facility——is the Prairie Village at Madison, S. D.

Prairie Village grew out of one of these Old Threshermen's groups.

After an Air Force career, Col. Joe Habeger came back to his old home town to teach mathematics at General Beadle State College, Madison, S. D. He won't talk much about his many bombing runs over Germany——just, "We were never shot down; just shot up."

Out in the shelterbelt on the Old Home Place stood a 1915 Case 20-40, "bought new." It had stood there for 25 years——even trees had grown up among its spokes, like the great oaks between Atlas' toes. Joe tinkered it back to life. Old Ralph Leonard, who had run that

Portable horse-drawn steam engine. Photo provided by John J. Menchhofer, who can say only that "It's an old one." Floodlights growing out of stack are not part of machine.

tractor for 30 years but hadn't heard it for 25, listened a moment. "She's runnin' about 65 RPM too fast, Joey," he said. Joe slapped his tachometer on it.

She was running 60 RPM over her rated 475.

Powering an old separator with the Case, Joe got some friends to help him thresh in 1961. This gave them the idea of an annual threshing bee, but they never got around to it in 1962. That winter, Joe and friends organized the Eastern South Dakota Threshermen's Association. The next year they got Adolph Rude to bring his Case steamer over from Flandreau, and some 6,000 people came to watch that one steam engine work on Joe's farm. The next year came 8,000 people during the three-day event, and in 1965, 10,000 people thronged to Joe's farm 13 miles north of Madison.

Hill climb used to be quite a thing at fairs——to see if you could get up the hill without losing steam. There's no draft at this angle. Case 50-HP; date unknown.——J. I. Case Co.

The Greater Madison Chamber of Commerce began thinking it was a pity these 10,000 people had to gather so far from the comforts of civilization. They voted it would be a kindness on the merchants' part to make the pleasures of their stores, cafes, service stations, etc., more accessible to the crowd, and suggested to ESDTA that they move their show closer to town next year.

Nobody will say who thought of it, but history will speculate it was Joe——somebody got the idea of recreating a turn-of-the century pioneer town to show folks how their forefathers lived while taming the raw prairie, and to stage the annual steam threshing jamboree at the village site. This grew into plans for an all-year-'round educational-recreational non-profit enterprise which could be used by tourists and scholars alike, where would be preserved some of our pioneer heritage. People started speaking of the idea as Prairie Village and the name stuck. ESDTA reorganized into the Prairie Historical Club, Inc. and elected Joe its president.

A church parading down the highway, jaunty bird's nest bonnet atop her steeple and followed by depot, ranch house and country bank headed for Prairie Village, got pictured in national and European papers.

Madison civic and service organizations jumped into a drive to raise $100,000 to buy land and build necessary shops and shelters so this irreplaceable equipment wouldn't have to sit out in the weather. Mrs. Irma Lyons made an ideally-situated 120-acre farm available to the club at far below market price because she liked the educational-recreational-preservation-restoration idea. Nestled between S. D. 34/U. S. 81 and Lake Herman, just two miles west of Madison. the land offered plenty of room for parking, displays, activities, the village itself and for growing the big grain crops required to keep a whole herd of classic threshing machines busy throught the three-day annual Steam Threshing Jamboree.

The project got not only local support but amazingly enthusiastic help from businessmen in Sioux Falls, the state's metropolis (pop. 55,000), 50 miles away, who contributed.

Volunteers pitched in and hauled turn-of-the-century buildings from a wide area: a church, a one-room country school, store, depot, country bank, farm and ranch homes, jail and a claim shanty. A photo the author took of the church marching down the highway, followed by bank, house and depot for the Madison Daily LEADER, was published in papers all over this country and in some European papers.

A log house was discovered near Canton, 100 miles away, built in the 1870's but clapboarded early so it was in perfect condition. Volunteers picked this two-story structure up and set it on the village site, where a tornado promptly scattered its logs over a wide area, mixed with pieces of barn.

Several "new" 1890 buildings are moved into Prairie Village each year. Traffic was backed up 2½ miles on Show Day.

Along with all their other work the volunteers rebuilt the barn and gathered up the logs. Hopefully, by the time you read this, the log house will be back in running order, for the tornado didn't hurt the logs any——just mixed them up.

Even before Prairie Village had a home, East River Electric Power Cooperative, Madison, was asking its 55,000 members throughout eastern South Dakota to donate or lend pre-electric or early electric household and farm gadgetry for a museum which it has since built at the Village. Dusty treasures flowed from attic and basement all around.

Those bundles don't just jump from the windrow into the shock--from shock to bundle wagon and then into the feeder. Somebody has to do the work. And every time they're moved, some of the grain shatters out. These were some of the sales pitches used in selling combines: less work, less wasted grain, no threshing crews to feed.

No, that's not a double-barrelled thresher. You see windstacker and weigher of a Case looming above and behind the Aultman–Taylor. Bundle-pitchers had to lay bundles in heads-first, had to keep machine up in the collar without slugging, and had to keep from falling into the clawing, chawing feeder because meat and bones gum up the cylinder.

The separator man was boss

There was always a running argument among early railroaders as to who was boss of the train——the conductor or the engineer. Legend has it that one conductor and one engineer settled the question for all railroaders and for all time with their fists: the conductor won.

The separator man is undisputed boss of the threshing crew. Everything centers around the threshing machine that it may do the best possible job. At 3 cents per bushel, wages and profits depend on getting the most bushels out in the shortest time. The engineman maybe was King in the old days——King in the sense of romance and glamour and prestige——but when the separator man says, "Jump," everybody still says, "How high?"

When he isn't oiling, inspecting and adjusting his machine and feeling bearings to make sure they aren't getting hot, he's standing atop his thresher, keeping an eyeball peeled.

Daring Young Lady of '74 shows some bare ankle at the Prairie Village old-time costume contest. Cheesecake not only caught the judge's eye, but the prize as well.

No, she doesn't have jets in her 1870 bonnet. Author snap-shot this from the hip as National Guard saluted Gov. Boe at 600 MPH during groundbreaking festivities.

What kids today have the enviable opportunity to jump in a bright, fresh strawstack? Just the kids who visit Prairie Village.

Fun for everyone at Prairie Village

The 3-day Steam Threshing Jamboree drew 13,000 people in 1966, the first year it was held at the village. You can get an argument going as to whether 20,000 or 30,000 attended the 1967 show. "Pretty good for a county of 15,000," grin boosters.

Not everybody registered, by a long shot, but those who did represented 39 states and 9 countries.

Two national-international publications sprang up to minister to steam buffs. Both represent enterprises that started as fun hobbies and grew almost to overwhelm their originators, much to their bemusement.

One is The Iron-Man Album Magazine, founded, edited and published by Rev. Elmer Ritzman, NTA chaplain, at Enola, Pa.

The other, Engineers & Engines, was started by T. H. Smith in 1955. Wrote a biographer, "Mr. Smith had spent most of his life working with steam engines on the farm, in oil fields, heavy construction and on the railroads." He died in 1961, but his widow, Mary Louise, continues the bimonthly, doing all the work including the printing, in her home at 611 Darcy Ave., Joliet, Ill. She has subscribers all over the U. S., in Canada and England.

A magazine devoted to old gas engine and tractor devotees recently split off one of these.

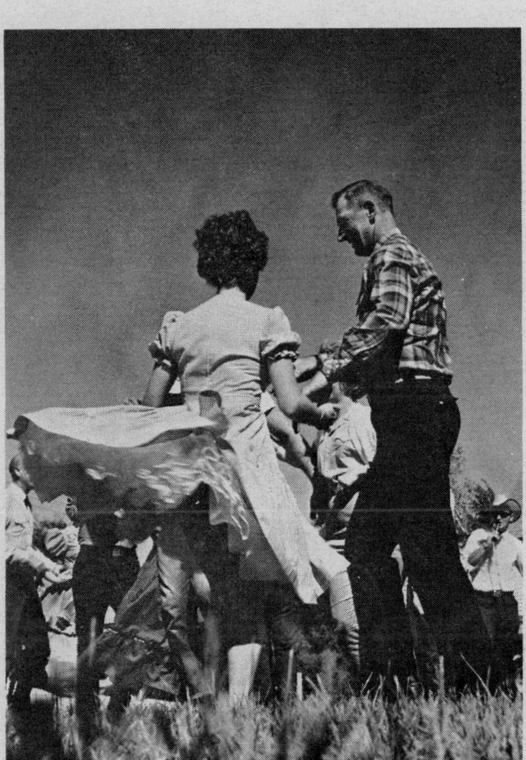

Squaredancers entertain the throng.

Not only lots of steam threshing and plowing going on, but auctions where you can buy everything from a lard press and a handcranked "labor-saving" wooden washing machine to a crankshaft for your '03 Goliath.

acres of models *Half-scale Avery thresher, bundle wagon and water wagon really thresh real grain. The only concession Garrett Havelaar, Hudson, S. D., makes is to open the bundles and hand-feed them.*

Have fun at Prairie Village

Prairie Village centers on steam-powered merry-go-round.

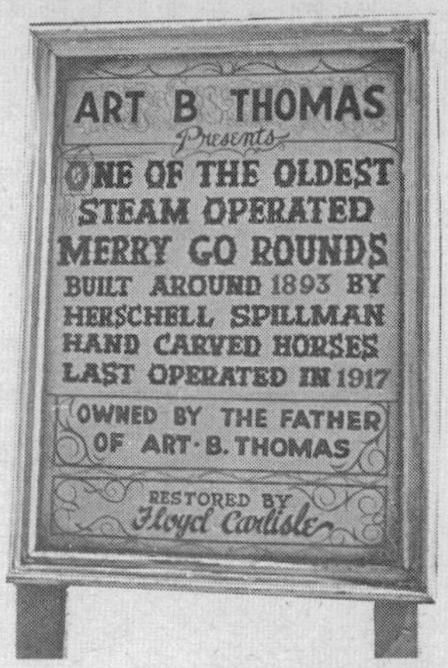

unique steam carrousel

Sam Johnson gives the kiddies pleasure with a sure hand on the whistle of the Herschel Spielman twin-cylinder opposed 8-HP engine.

When Herb and Mrs. Willoughby bring their collection over from Howard, they make quite a parade. They're seated in a 1920 4-door Franklin.

Little Kids growed big have a ball operating the real coal-burning engines from 1/10 to 1 HP they built themselves. Every show features 1 acres creeping and crawling with real operating model. Leo Huston, Watertown, S. D., built this ¼-scale approximating the Nichols & Sheppard.

CASE

Steaming up

We would not presume to tell an old hand how to fire up––you fellows were experts ere we were even a gleam in Daddy's eye. Nor are we trying to give the city slicker a short course in steam engineering. But we thought folks who've never fired a steamer might like to know there's more to it than just turning a key. Our mentor here is an experienced engine man who wishes to remain anonymous lest the neighbors think he's trying to tell them what to do.

Biggest day in the spring for a hooked steam man is the day he first fires up. It's an impatient wait during the lengthening days as winter reluctantly wanes. He watches the weather reports and goes to bed with one eye on the sky, hoping for a mild day tomorrow. If the air's below 22 degrees, the water will freeze before he can get it hot.

Meanwhile, out in the shed he's looked the boiler over for wet seams and flue tubes. [Most farm steam engines had fire-tube boilers: the 36 to 76 flues (depending on the size of the engine) are 2-inch seamless ductile steel pipes carrying hot gases and smoke from the firebox forward, through the water, to the smokebox and out the stack. This design is called a fire-tube boiler: the tubes carry the fire. This design was characteristic of the railroad steam locomotive. (Indeed, the old steamers today are repaired with used flues removed from railroad engines.) Big power plants, steamships, etc., have water-tube boilers: the tubes carry water, surrounded by fire. (See diagram, p. 44)

A white crust around the tube at the endplate indicates she's leaking where the flue fits into the endplate (a rolled fit, not welded). A few simple minutes with roller and hammer snugs them up. If the white is inside the flue, this indicates a leaking flue which must be cut out and replaced. When the trouble is at the end, the tube can be tipped––a new end section welded on.

In cleaning flues with his spiral-steel reamer on the end of a 12-foot rod, any extra resistance the cleaner meets indicates the possibility of a split.

Contrary to what you might expect, a leaking flue does not explode. It just leaks. Water drips out the smokebox and, if the leak is real bad, out the firebox.

Flues aren't too hard to find. New ones are kind of expensive but for $5 you can pick up used flues from locomotive, school and apartment house boilers. Replaced in routine maintenance, these used tubes have got 20 or 30 years left in them for an engine that's seldom used except at annual shows.

A white crust around a flue tube indicates a leak.

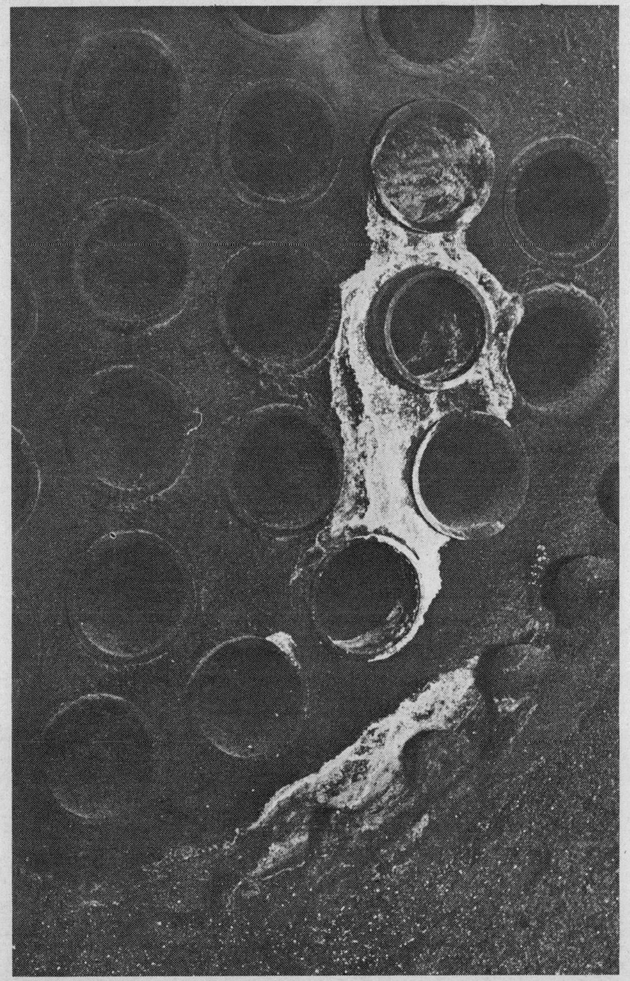

If the cleaner meets resistance inside the flue, that may indicate a split.

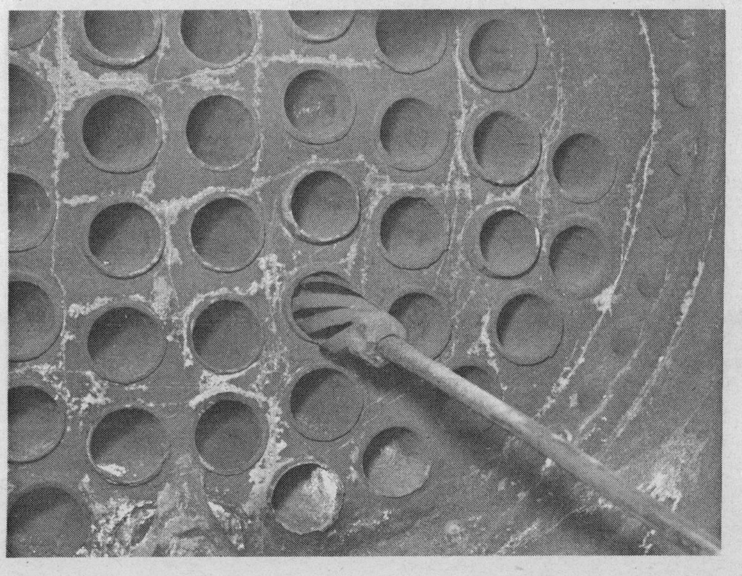

He cleans the flues––a job that has to be done several times a day, especially if the load is light. Heavy going blows some of the soot out.

Bottom left: Long-handled tube cutter reaches through smokebox, cuts end off defective flue. Shorthandled cutter is used in firebox.

Upper left: Split flue is pulled out through front flue plate, new tube inserted. Firebox end is beaded over to keep fire from burning ends off; bot ends are rolled tight.

Upper right: Tubes are roll-fitted into the endplates or flue plates... no welding, no gasketry, no cement here. This is the tube roller: hit it and turn it and hit it again.

Lower right: Rolling tubes is a dirty job if there's a fire in her.

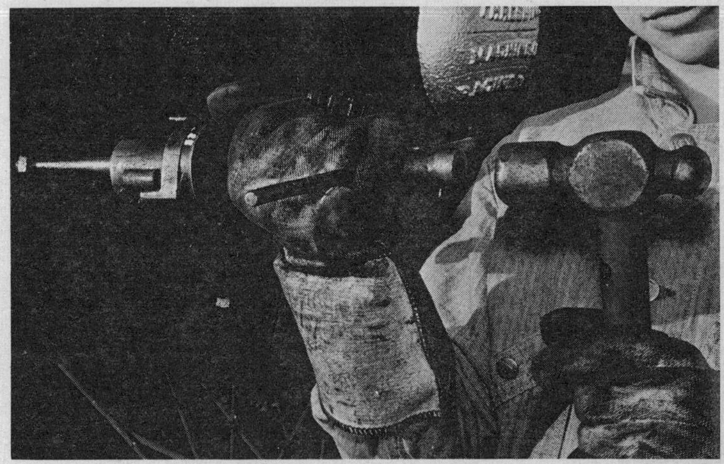

To replace a flue, you cut off both ends, pull her out through the smokebox, push the new one in. You crawl into the smokebox and bell over the flue end so the fire won't burn it off, then roll it tight. Roll the smokebox end. It isn't too much of a job when you know how and have the special tools, but hard on the ear drums. A pot-belly is no advantage in the firebox.

Be sure the fire's out.

Father and son, heads wrapped to keep soot out of hair, make a tight pair while rolling tubes inside firebox, which in this engine is 42 inches long, 30½ wide, 27 high. Dad's showing son how to "Roll Them Tubes." Camera is looking through inspection hole with Dad holding flashgun inside firebox.

While waiting for the weather to warm up he's caulked the seams by beating them tight with a cold chisel.

Our engineer friend checked the grates and chipped off any clinkers building on them. Clinkers slow the airflow, producing a smoky, poky fire and burning out the grates because the incoming draft keeps the grates from overheating.

It takes 250 gallons to fill a 60-hp boiler——a ton of it. He's careful to keep oil, detergents and soaps out of the tanks he hauls water in, for they may promote foaming and priming.

"If you won't drink it, don't put it in your boiler," is one old engineer's counsel. In fact, the human stomach is far better equipped to handle impure water than is the boiler. Mineral from hard water forms scale (like in your teakettle) which not only insulates the tubes and cuts engine efficiency——when the engine cools, the scale shrinks more than the tubes and squeezes them loose from the endplates.

Our engineer prefers rainwater from a cistern. If he can't get it, he bails it out of a dugout or pond. He's even been seen hauling it from the water softener company in town.

He doesn't believe in boiler compounds to reduce scale. As an engineer on a steam-powered seagoing tug during WW II he learned, "The more junk you put into a boiler, the easier it will prime and foam."

Two levels in the water glass are of life-and-death importance to the engineer. One is the level at which water just comes over the crown sheet (the "roof" of the firebox), and the other is an inch of water over the crown sheet. When he's firing a boiler he's never fired before, he takes out the wash-out plug or hand hole plug above or in the back of the boiler so he can see the crown sheet. While filling the boiler he keeps a close watch through the hole and notes the water column level when water first wets the crown sheet.

The crown sheet is his life insurance: if he lets it get dry, it gets red hot and can blow out, killing him and everybody in sight.

He fills the boiler until the water glass is half full.

Into the firebox he throws oily wiping rags, then straw, cobs or kindling-wood. He may throw in a little kerosene or diesel fuel (NOT gasoline!!!). When he gets a thin, even bed of red-hot coals he starts putting in coal. Engineers used to buy egg-size nut coal, but nowadays it's hard enough to find coal of any kind and you have to unload it yourself.

He has his fireman or flunky break the big chunks into fist-, egg-, and nut-size pieces: by holding the chunk in one gloved hand and hitting it with a hammer...not just for firing up but for the whole run. Uniform small-size coal makes easier firing.

"If you wouldn't drink the water, don't put it in your engine," say the Old Timers. Hose comes from barrel pump on tankwagon. Takes 250 gallons to fill this boiler.

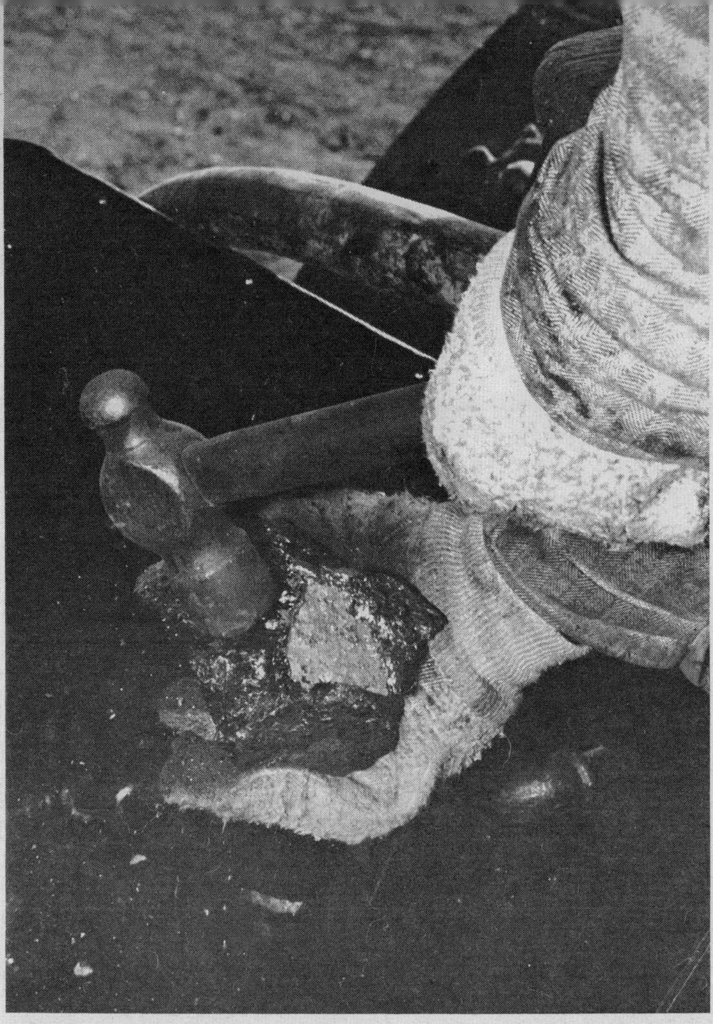

If he can't buy egg or nut coal, he has to break it in his hands--one lump at a time.

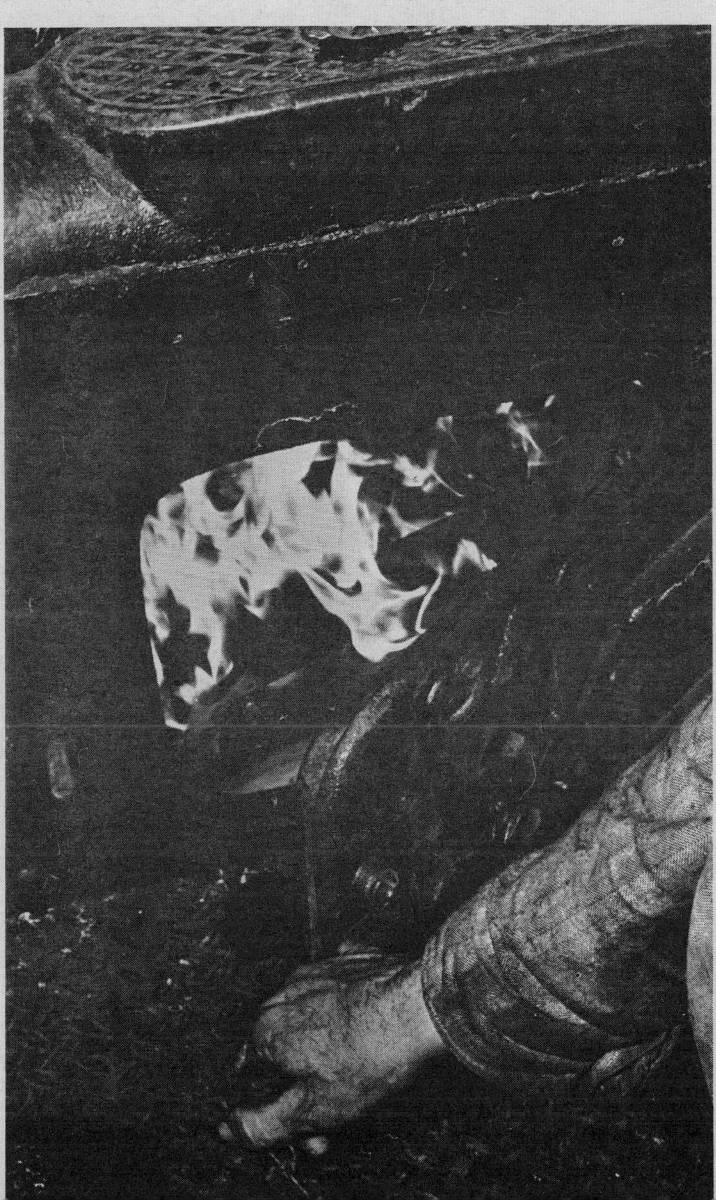

Oily rags, straw, cobs and/or kindling wood start the fire. When he has a good bed of red-hot embers, he lays in a thin layer of egg-size coal.

Getting up steam

He keeps the fire even and spread to all corners of the firebox, level all the way across—–not heaped in the middle. He keeps the fire door closed except for throwing in fuel. Occasionally he pokes the coal with the shovel turned over. This is all the stirring his fire needs while building pressure.

The boiler goes through a sweat: water vapor in the combustion gases condenses on the water-cooled tubes until it drips out fore and aft. You can't tell a leaking tube from sweat at this point. The sweat will dry up as the boiler heats; the leak will get worse as pressure builds.

When she starts to boil she sings like a teakettle. While the boiler is warming up, the engineer and his crew thoroughly oil the engine and fill the hard oilers with hardoil (what I always thought was filling the grease cups with grease).

He watches the hand hole and wash-out hole filler plugs, snugging them up a little if they leak.

First-steamed water may make 15 pounds pressure in a surprisingly short time. This is just air driven out of the water. The stack blower soon bleeds this off.

The stack of a farm steam traction engine, compared to the powerhouse smokestack or even your house chimney, is relatively short and therefore doesn't produce much draft. A steam jet in the stack creates more draft on the venturi principle. A good engineer, however, doesn't need the blower much to raise steam. "He'll get up early enough to get his fire hot in time for threshing right after breakfast," says an experienced threshing engineer.

When steam gets up to 25 pounds, he checks all the hand holes and draws them up snug. "Don't use brute force on the hand hole bolts," he warns, "because it isn't needed and you'll only ruin the frog. The same applies to the brass wash-out plugs. Brute force here a few times and they'll have to be replaced because they cut away. You may strip the filler plug in time.

Installing A Never Leak Gasket

"If the hand holes weep, they'll quit as more pressure builds. If they leak, the leak will get worse with pressure. Then you'll have to cool and drain the boiler and replace that hand hole gasket."

If a hand hole gasket leaks, he cools and drains the boiler, takes out the hand hole plugs (holes for washing the inside of the boiler) and cleans the holes thoroughly, removing all the old gasket and any scale. He does the same with the hand hole plate.

He oils the bolt and runs the nut up and down the bolt with his fingers until the nut goes on easily.

Using an old gasket for a pattern, he cuts a new one out of good gray or red rubber low-pressure gasket material, making the outside just a little larger than the hand hole plate. He cuts the inside hole just large enough to fit snugly on the hand hole plate.

After getting the hand hole plate and gasket back into the hand hole, he straightens it up square with the hole. He puts the frog on the bolt and runs the nut up finger-tight. Now he eyeballs hole, plate and gasket.

"You should be able to see a small amount of gasket between the plate and the hand hole. If you see the plate is going to touch the hole, move it slightly so it won't touch when you draw it up. Don't be mean—–just tight enough so it won't leak. When you've got 25 pounds of steam, tighten it again. It'll never leak, although it may weep a little."

But an engine's tears, like a woman's, with tender loving care are soon dried.

After he gets the leaks stopped and the boiler stops sweating and starts singing, he opens the smokebox door at the front of the boiler and pokes the flues again.

"Do a good job fast," cautions our engineer. "You're wasting heat and cooling off the boiler. When I was small I could never do it fast enough for Dad. He always finished it when I was half through."

Closing the smokebox door, he goes back to the platform and opens the draft halfway and opens the blower. "Watch the soot come out!" he chortles. "But don t do this with a crowd of well-dressed people around. They won't like the color of their clothes. Then shut off the blower and begin to warm up your engine. This isn't easy, either. Be careful."

While the boiler's warming up, he fills the oilers

...and fills the hardoil cups with hard oil.

Wheels...

Crosshead slide...

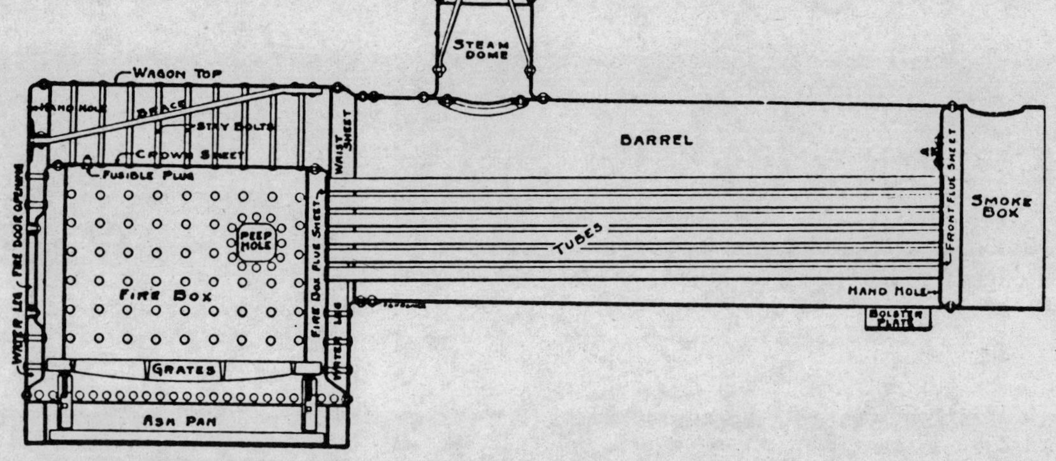

SECTIONAL VIEW OF BOILER--COAL AND WOOD BURNER

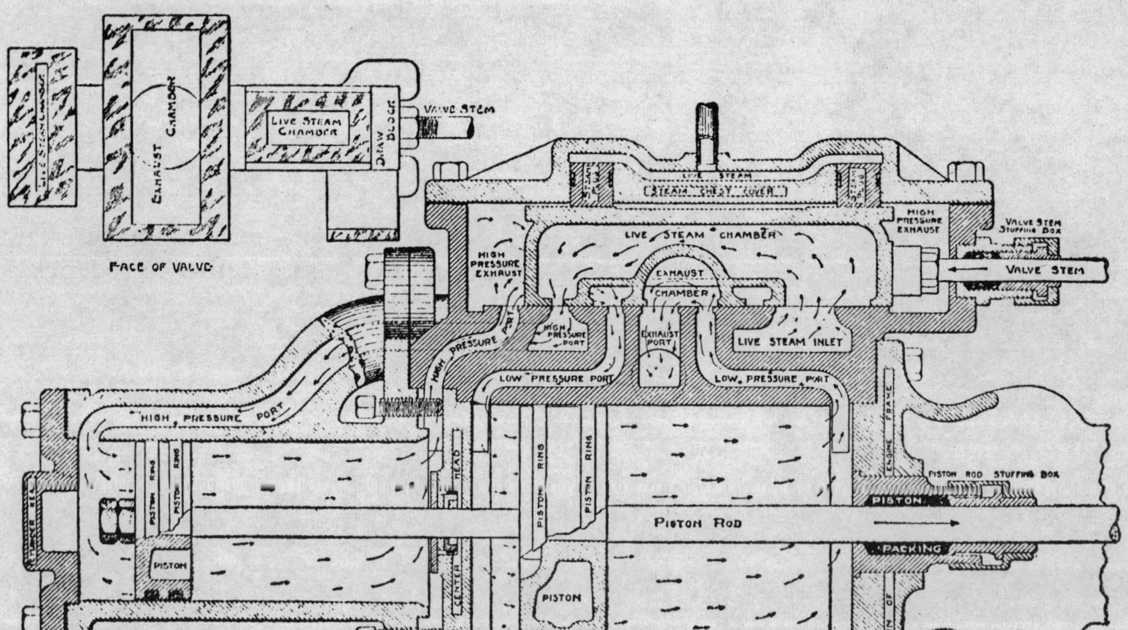

SECTIONAL VIEW OF WOOLF COMPOUND CYLINDER

Engine was 4 years old when this picture was taken in 1921. Proud engineman, Fred Klinkner, Scotland, S. D., is third from left.-- Photo courtesy George Klinkner

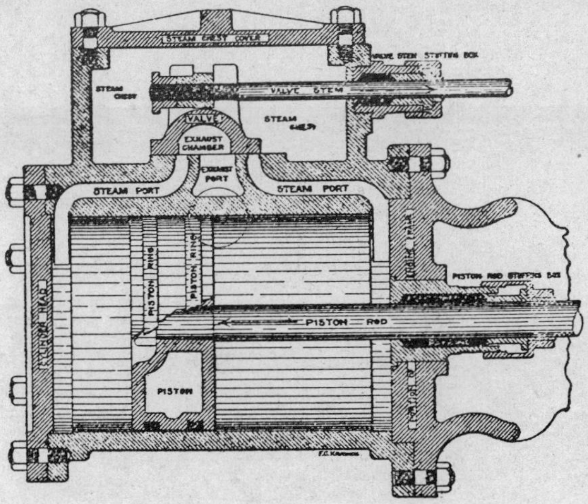

SECTIONAL VIEW OF SIMPLE CYLINDER

Engine" sometimes refers to the entire machine, sometimes to the boiler alone, but properly to the engine proper: the cylinder, piston, flywheel and attachments.

Most farm steam traction engines were single-cylinder "simple" engines, although a few were two-cylinder and some were "compounded." In a simple engine, steam expands once in the cylinder and then exhausts through the stack. In a compounded engine, steam expands first in a small cylinder, then again into a larger cylinder (both on the same push rod), where it does another batch of work before escaping into the stack.

The J. I. Case Threshing Machine Co., leading steam engine maker, didn't think much of compounded engines, apparently, although its 1908 catalog offered them in 27, 36, 45, 60 and 75-HP. The catalog commented, "In some sections and for some lines of work, the 'compounded' has a slight advantage over the 'simple'."

In the 60-HP engine, for example, the simple engine had a 10-inch bore and a 10-inch stroke; the compounded had a 7-3/4-inch high-pressure cylinder bore and 11-inch low-pressure bore and a 10-inch stroke. It cost $100 more. Both operated at 130 pounds.

The engine posing for the photos in this section is a 60-HP 1917 Case simple single cylinder, 10-inch bore, 10-inch stroke, 250 rpm, 40x12-inch flywheel,, 30½-inch barrel (boiler) diameter, 46 flues 2x90½ inches, firebox 42x27x30½ inches, grate area 7.88 square feet. It had 216.9 square feet heating surface above the grates, carried 1,540 pounds of water at working level, a ton including feed water in the tender. Working pressure is 140 pounds per square inch. It had one speed forward, 2.61 mph, and the same in reverse. Cost $1,700 new. This engine will run on wood, coal or straw; for burning straw, a straw chute and firebrick arch were added, as well as a wood jacket. It is equipped with the jacket.

It is owned by George Klinkner, farmer, Artesian, S. D., whose father, Frederick, bought it new in 1919.

Starting the engine

At about 50 pounds steam, he very carefully opens the main valve on the steam dome, then the valve to the steam chest

When pressure hits 30 to 60 pounds, depending on the size and make of the engine, our engineer opens the main valve on the steam dome very slowly. If he gave it a whirl, a sudden steam blast could carry a slug of water that would hit the pipes like a shot shell and bust them. Valve open, he just barely cracks the throttle

With the throttle opened slightly, he opens the water bleed-off on the steam chest——the box-like affair just above the cylinder which houses the valve and where the steam is shunted to one end of the cylinder and then the other. When steam comes through the bleed-off and has warmed up the steam chest, he opens the cylinder cocks and starts rocking the engine with the reversing lever, rocking it until the cylinder is thoroughly warmed and all the water is worked out of it before he allows a complete revolution. Water won't compress, you know, and if there's enough condensate in the cylinder to fill the space between the piston and the cylinder head, a complete revolution would break a piston or blow the head. For this reason, whenever he stops the engine, he opens the cylinder cocks to let the water out before starting it again. (Some engines have automatic cylinder cocks, similar to steam traps.)

The first engineer who gets up enough steam to blow the whistle in the morning scores a point on all his neighbors. In the old days, engineers put as much store by this as the housewife did in being the first to hang out Monday's wash. As each engine man got up pressure, he tooted to let the neighbors know he was in business. Shrill shrieks wafting across the prairie made the sun come up most musically. Whistles were a status symbol: everybody wanted a bigger one than everybody else.

He caresses the throttle with a lover's hand so as not to jar the gears. These are cast iron, brittle as glass. Open, they're lubricated with axle grease and dust.

The engineman watches the sight feed to make sure the steam chest oiler is working, never taking it for granted.

He keeps an even, thin fire, not a large lumpy one. He keeps ashes in the ash pan from piling up to touch the grates, for this will burn the grates—again, by blocking off the cooling draft.

He keeps the grates well-covered with a thin layer of hot coals. Whenever he throws in coal he checks his fire for holes and fills them. "You can easily see the holes in your fire: the grates show up. Gray smoke indicates holes in your fire." He spreads his coal as he scoops it in. "Don't keep the firebox door open too long as this lets cold air in and cools the boiler fast."

Engineers and photographers live at opposite ends of the universe. They've forever got a running battle going on. The good engineer works for a clean exhaust, "Just a whisper of smoke," but the photographer wants vast billows, the blackest possible. Most engineers will, however, as a special favor, throw in a big splash of coal, some straw or even an old tire to make smoke for a picture, even though it hurts the engineer's pride, his flues, and—worst of all—his reputation.

"Sorry about that," says the photographer. "Toss in another tire."

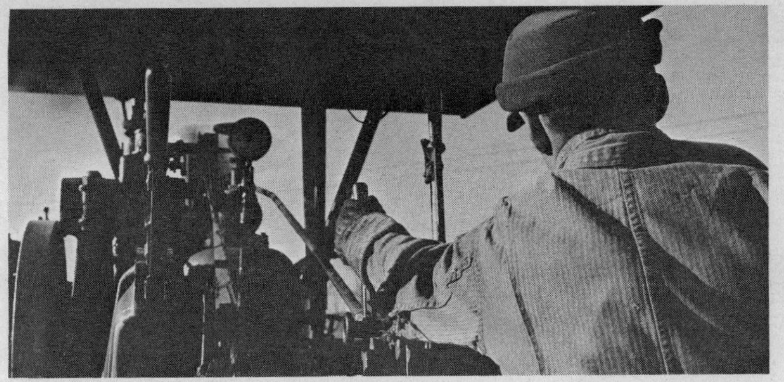

He cautiously cracks the throttle.

With the reversing lever, cylinder cocks open, he rocks the engine forward and back, working the water out of the cylinder.

Injecting cold water cools off boiler, wastes fuel. Pre-heater (cylinder ejecting steam) warms feed water with exhaust.

Warming up is touchy business. Note intense concentration.

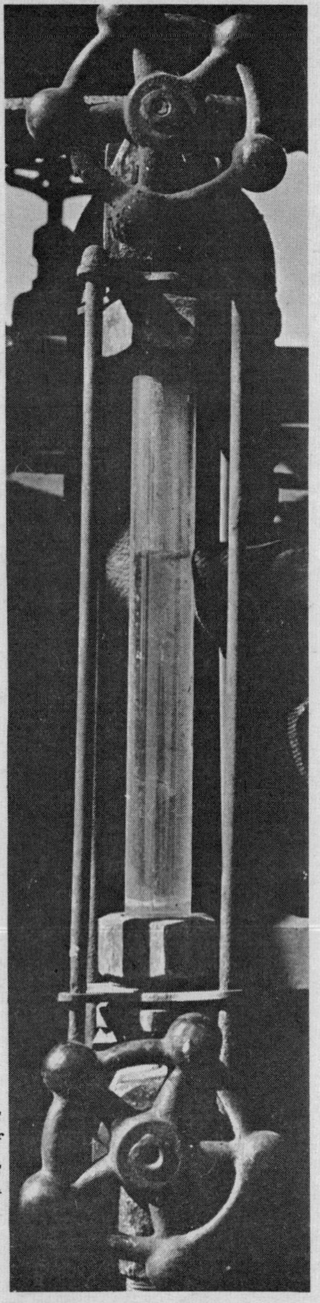

Engineer watches water glass as if his life depends upon it––which it does!

The engineer keeps an eyeball peeled on that water glass. He tries to keep water at least an inch above the crown sheet for safety.

"The best firing is when the water is between one inch above the crown sheet to half a glass, depending on the load. In easy going you don't need so much water," says our engineer.

The engine carries a 350-gallon tender to replace water evaporated into steam and blown out the exhaust. When there's pressure in the boiler, the engineer turns on the steam-powered injector which forces water in against steam pressure. Since the injector is not automatic, it's the engineer's skilled hand that keeps the injector working at just the right speed to keep water high enough for safety and low enough that she won't prime.

Pulling the engine hard, demanding more steam than the boiler is producing with more than half a glass of water, may make the boiler prime: water goes over with the steam. This will make the engine knock and can, as we described, ruin the engine. When she starts to clank, he opens those cylinder cocks before you can say steam traction engine. "Closing down the throttle will usually stop the boiler from priming."

When he gets the water worked out of the cylinder, he closes the throttle until the priming stops. He then checks the fire for holes and breaks up any large lumps, fast. If he has a lot of unburned coal in the firebox after the engine has primed, he stirs the fire a little. If he's been using too much draft, he closes it down half, throttles down the water injector or shuts it off.

"The engineer watches the water column as he works back to his load, working the water a half to an inch-and-a-half below its previous level," reveals our engineer.

Cutting the pressure and opening the throttle wide increases the tendency to prime. An even, hot fire without holes and a controlled draft help control priming. "With a hot fire and just enough water, opening the throttle only a fourth to half an inch more than needed for governor control is safest," says the engineer.

Foaming is different from priming. Foaming produces bubbles not only on top of the water but all through it, raising the foam-and-water level maybe into the steam dome, which is there to let water fall out of the steam so it won't be carried over into the engine. Modern dish and laundry detergents are getting back into our water supply and increasing its foaming tendency.

Takes a quick hand on the throttle (center lever) if she begins to prime. Right-hand lever is clutch; left-hand, reversing gear.

Maybe she's got a lumpy fire. He stirs with his shovel turned over.

He's working with 100 to 155 pounds per square inch steam pressure here. (Most farm steam engines were designed to work at 125 to 135 pounds; some state laws limit this to 100.) While the boiler was built with a safety factor of five and factory-tested at 225 pounds, that was 50 or 60 years ago. And that safety factor is figured on a boiler that isn't red-hot. When water gets below the crown sheet all rules are off: a red-hot crown sheet will buckle or blow at working pressure.

Water drops fast as it gets close to the one-inch above the crown sheet. This isn't much of a margin on rough land, for on a down-grade the water rolls to the front of the boiler, leaving the crown sheet high and dry. Should the engine head downhill long enough for the crown sheet to overheat, then start up-hill, cascading water onto this thick chunk of red-hot steel, the gunpowder workers' wives' prophetic song would come instantly true for the engineman's widow:

Someday there'll be a thunder sound
And scattered far and near
O'er hill and dale and all around
Will be my husband dear.

If his pressure is high and he's dangerously low on water, he stops and takes on water.

"It's better to stop than to ruin a boiler.

"To use your injector, open the steam valve on the injector wide and throttle the water side until the water doesn't go out the overflow. Sometimes an injector gets hot and won't draw water. Cool it by pouring water on it and take off the suction hose and let the hot water out. Keep a cool head and don't get excited. Good injectors do work."

Some state laws required two injectors or water pumps.

If water drops below the wet crown level, the engineer "Is going to be a busy man. He throws the belt if he's threshing, pulls the pin if he's plowing, and gets those front wheels up on some bundles or backs the drive wheels into a deadfurrow. Then he goes to work on that water."

There's a fusible plug in the crown sheet that melts before the crown sheet gets dangerously hot, letting steam into the firebox to smother the fire. But sometimes the plug scales over so heavily that, even though the plug melts, the steam can't blow through and smother the fire like it's supposed to.

Even if the crown sheet doesn't blow, overheating can warp it so it tears loose of the staybolts and ruins the boiler.

"If the fusible plug melts, there's nothing for it but to pull your fire, cool down the boiler, crawl into the firebox and replace the plug."

The surest way to a short life is to inject water onto a red-hot crown sheet. This five-eighths-inch-thick steel holds a lot of heat, makes flash steam. Flash steam is uncontrollable and makes that thunder sound, carrying to your widow news of her late husband.

"If you have any doubts about the crown sheet temperature, just stick your shovel into the firebox. A red-hot crown sheet will throw a bright red glow on the shovel." Red signals danger. When you see red, here's your Standard Operating Procedure:

"Open the safety valve and hang the poker on the handle to keep it open. Close your draft and dump your fire. Stay there and try to cool off your boiler. Don't run, or a hunk of boiler will catch you."

Many operators today run at lower pressures because, after all, these boilers aren't as young as they were 60 years ago. Rust and corrosion could have eaten a thin spot. And since steamers are used just for show these days, they don't have to work so hard. The less hard they pull the longer they last. One operator refuses to plow with his steamer because the heavy drawbar load cuts out the gears. And when the gears are gone, there ain't no more nohow. Today's farm implement dealer is fresh out of steamer parts.

If the crown sheet gets red, he hangs the poker on the safety valve and starts pulling his fire. To run is to die.

Shutting her down

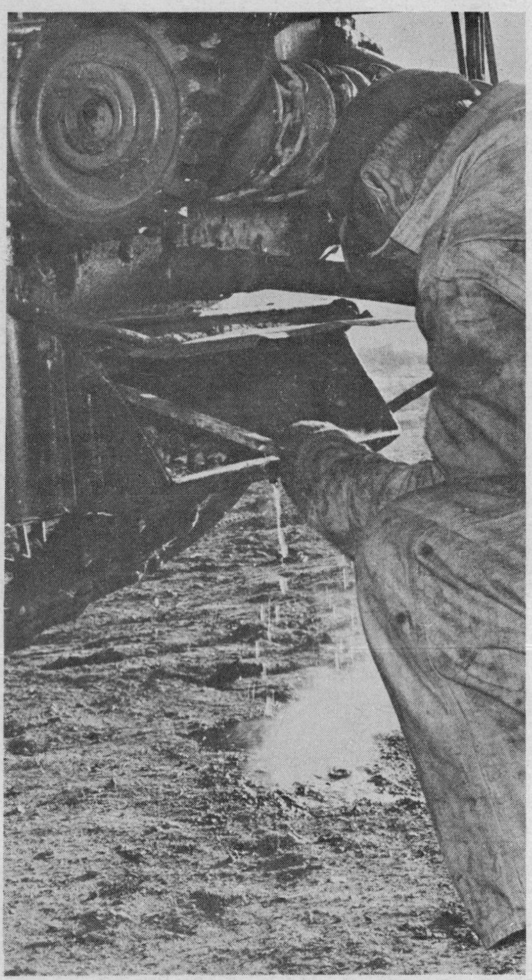
You have to pull the fire, too, if a leak develops, and to shut her down for the night.

When his day's work is done the engineer isn't through yet. There's more to shutting off a steam engine than just flipping the ignition switch.

He dumps his fire and cleans the ash pan, scraping the ashes out of the corners. Ashes are hard on the water leg and water bottom. Ash draws moisture; moisture activates the sulfuric and other acids in the ash, causing severe corrosion.

When he shuts her down for the night, he closes off the main steam valve by hand. "A wrench will damage the seat." If the engine will run again tomorrow, he takes on enough water to have at least half a column (in summer; if there's the least chance of freezing, he drains her all over).

He lets the fire burn out, but he closes the drafts so the boiler will cool slowly. Although it takes two hours to get up steam in the morning, he doesn't try to bank the fire to keep steam up because this is hard on that part of the boiler and will cause the grates to heave under the fire. He just gets up a couple of hours earlier.

If the engine isn't going to be used again for a while, and ESPECIALLY if there's freezing weather in prospect, he drains her completely––opening not only the main drain but also the cylinder cocks, steam chest bleed-off, tender drain and disconnecting all pipe unions.

The most important thing about shutting down the engine is forecasting the wind's direction on the day he'll fire up again, whether tomorrow or next Spring: the wind must blow right into the firebox door. If it doesn't, the draft will be slow, the fire balky, and it may take four hours instead of two to get steam up. And you don't move a cold engine around very handily––takes a whole herd of gas tractors to budge it.

Then when the fire is out, he caps the stack.

If he forgets to drain anything, it'll freeze up and bust. Here he drains the throttle.

Main drain empties boiler for winter. Engineer can also blow out sludge here.

ıjector line gets its plug pulled, too. Here you get a good view of the primitive orm-and-chain steering gear.

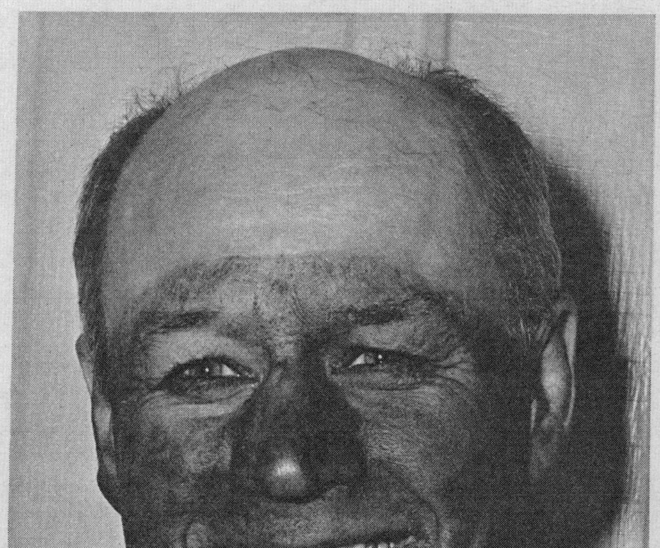

Portrait of a Happy Engineman (you can tell by the soot on his teeth). But Mom, strange creature, prefers gas tractors.

He caps his work by capping the stack.

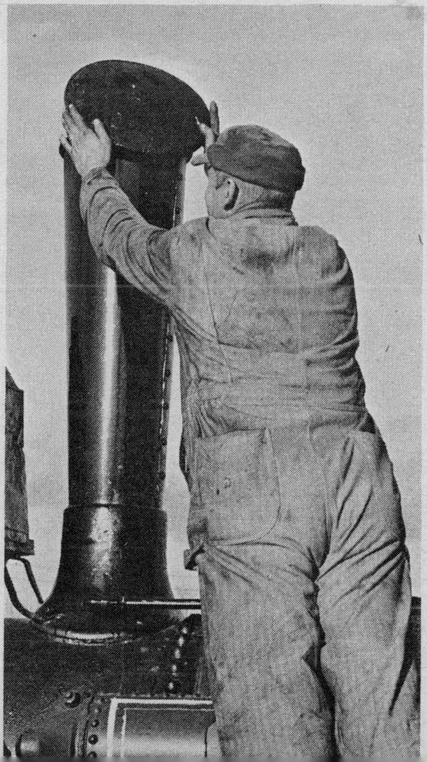

The steam engine man a dying race? . . . . . .

The lusty steamer man of 1920 is gray today. When his breed is gone, there will never be another and his lore will be interred with his bones.

Or will it?

There's a new breed of engineers coming along. Though their chins are yet as bare as their elder's pates, their eager young minds are soaking up all the old-timers can tell them and show them on the care and operation of steam engines. The love is in their hearts unbidden.

One of these was a steam engine man at age 11. Today, at 14, he can bead flues, clean tubes, fire up, belt up to the thresher and keep the old brute singing a busy song right along with engineers of 50 years' seniority.

His grandfather Fred Klinkner paid $1,700 for a new 1917 60-HP Case steam traction engine and $695 for a 44-inch separator. He threshed for years around the home farm at Scotland, S. D. Granddad taught his son, George, 10 years younger than the engine, all he knew about steamers. But Granddad Fred passed on and George went to war. For a third of a century the old steamer sat in the shed, cold and silent. It's a wonder she didn't go overseas as shrapnel——the fate of many a brave engine.

A proud day——after months of crawling in and out of the firebox, rolling tubes, greasing bearings, oiling the cylinder, breaking coal, hauling water——Donald Klinkner at last graduated to the controls of Father George's 1917 60-HP Case steamer to become, at 12, the world's youngest steam engineer.

Don started learning the steam engine business at age 11. He helped his father restore the aged monster to mint condition.

. . . . noises from the grave

Then in the early '60's came noises from the grave: steam was reborn——not for getting farm work done any more——two hours steaming up isn't like turning the starter switch on a modern tractor——but as the biggest, lustiest, heaviest, dirtiest and, its disciples assure you, the funnest hobby going.

When the eastern South Dakota Threshermen's Association started staging annual steam threshing bees with a single borrowed steamer on Prof. Joe Habeger's farm near Madison in 1963, George yoked three modern farm tractors together and hauled his dad's old Case out of the barn. He was farming at Artesian, 60 miles west of Madison, by then, but distance and hiring a 20-ton lowboy was no matter. Here was a chance to bring the sleeping beauty to life, all 10½ tons of her muscle, whistle and affection!

He ransacked the attic for the owner's manual, scrounged around the archives of his brain for near-forgotten lore learned as a kid, reamed the flues and fired her up. The piston, that hadn't moved for a third of a century flashed. The dusty flywheel turned at last, throwing off clouds of chaff and spiders.

And 11-year-old Donald was right there, eyes big and bright. Before his next birthday he was the world's youngest steam engineer.

Ever since he was a little fella just big enough to clamber into the cab, Don had scrambled around over the boiler, explored the controls, rattled the shovel in the cold black firebox, going choochoochoo as he and the old girl conquered childish worlds. He didn't have to learn to love the monster. They'd been pals since before he could remember. And it wasn't much of a problem to learn to clean and fire and run her, either. With Dad as his teacher, his mind soaked in instructions like bread sops up milk.

While his city cousins are entranced by rockets and satellites, while some of their neighbor kids go on rumbles, young Donald is scouring flues, restoring the old beast, running it or is folded up inside the firebox beading-over flues.

By the time he was 12, Don was good enough to run the steamer in the Madison shows. Dad had no qualms about leaving him in charge despite the awesome responsibility. With a huge crowd gathered 'round like that, a careless engineer could kill not only himself but a hundred others.

Don likes it all. Even the dirty work.

Ask him, "What's your favorite sport?" and you get a quick answer:

"Steam threshing."

"Your favorite work?"

"Working on the Case."

"Your favorite music?"

"The steam whistle."

"Your favorite perfume?"

"Coal smoke and Mom's cookin'!"

An engine, protected from the weather and kept oiled and in repair, will last forever. Boiler plate can still be rolled. Tubes can be picked up in junkyards. Bearings can be poured from a ladle of molten Babbitt metal, a sheet of paper wrapped around the shaft for clearance. Worn-out gears can be rebuilt with the welding torch, or can provide the patterns for new castings.

Steam engines are immortal. And if each new generation of enginemen begets another, and if people continue to crave new experiences, crowding around these gentle giants from the past, the whistle's joyous shout will roll across many a prairie yet.

Long months of instruction preceded his laying a hand on the controls. Dad shows him what the levers are for. First thing he learned to operate was the whistle, and that's still his favorite, Mom says.

small part of the crowd that gathered to watch Don thresh wheat at the 1967 Madison, S. D., Prairie Village Steam Threshing Jamboree. The gineer has a profound obligation with this many people around to keep anything from happening.

Nobody's down below except a Governor, a U. S. Senator, a Queen and some town leaders. Up above, the Engineer is King, Seer and Hob-Nobber with the gods!

You're a happy man when you get that throttle in your hand, the engine throbbing, the stack barking, the road unwinding behind you at 2.61 MPH!

Why a steamer is a dreamer

There's an air about a steam engine man.

I don't mean just the fragrance of coal smoke and crankpin oil that clings to him months after he's drained her for winter. I mean that far-away look of the prophet who sees things unseen by mortal eyes––the aura of mystery surrounding the great soul who hobnobs with gods and is privy to secrets closed to mere earthlings.

For the steam engineer is master of power such as turns the world, that unlocked Earth's cornucopia. Under his cunning hands this strength, equal to fourscore or more great horses, meekly serves humankind, issuing from a black tube as mysterious and potent, in its way, as the nuclear physicist's black box.

In that sooty head is lore the ancients never dreamed of––how to get steam up in hardly two hours––what black smoke means, what gray smoke means, how to prevent holes in your fire––how to keep the flues clean––when to set the cut-off for heavy load, how to slice it for light––and even how to keep the boiler from blowing up.

Engineers are a bold race, and proud, men the world has passed by, much to its loss.

The Day I Drove a Steamer

Speed Kills.

You can get into trouble at 1/8 MPH.

I asked our friendly neighborhood engineer if I could drive his 1917 60-HP Case. He said sure. (He'd already taught me how to blow the whistle.)

First of all, there's a trick to steering. The old worm-and-chain steering gear has a ratio of about 40:1. It's practically impossible for one man to turn the wheel when she's settin' still. But when she's moving, no matter how slowly, even I could spin the wheel with one hand.

I was under a bit of a handicap what with two cameras and a 10-pound flash powerpack around my neck and a camera in the other hand. But I soon discovered you have to anticipate your turns about a quarter of a block ere the corner and get that steering wheel to spinning. It takes two full turns just to crank the slack out.

Not knowing whether the throttle opened forward or back was a trouble. Not knowing whether the valve lever went whichaway to reverse the engine, and forgetting all about the clutch lever, I think I made the engineer's bald spot turn gray as I almost hooked a threshing machine, skinned by a 10-ton 1914 Minneapolis Farm Motor and rounded the corner on four wheels into the lane without quite hitting anybody. When you don't know what you're suppose to do nor how to do it, a snail's pace is the pace of fools.

But I'll tell you, man, when you get those 10½ tons zooming along at a full-throttle 2.61 MPH and feel that invincible surge of mighty power beneath your hand and the rhythmic heave of the steel deckplates under your feet and hear the bark of the exhaust in the stack and feel those cinders in your eyes and taste the soot upon your lips and suck in the heady, hot cylinder oil––man, you're hooked.

There's the soddy--and in the foreground is the one-horse breaking plow a pioneer plowed up his sod "bricks" with--and that plowed the sod strips for this soddy. Claim shanty is in background at the Madison, S. D., Prairie Village.

Where you can see how your forefathers lived

Back East, where Nature provided lots of trees, the glaciers left plenty of rocks. Settlers built log cabins or stone houses. Building materials lay at their feet. Fuel was in the forests, in David Thoreau's words in WALDEN, "an encumbrance."

Not so on the Great Plains.

An ancient sea floor, the Plains provided neither trees nor stones, but still homes and fuel lay at the ingenious pioneers' feet, though less easy to come by. Many of them, too poor to buy and freight lumber from the railhead maybe a hundred miles away, built their first homes with the only thing at hand: the tough, matted prairie sod.

An ox team and a breaking plow turned the sod in foot-wide strips about four inches deep. The farmer, his family and friends cut the half-mile-long furrow strips into manageable lengths of two or three feet and laid them like big bricks. The tool of choice for cutting was a corn knife, but some used a butcher knife. Moderns, finding corn knives scarce, use a linoleum knife.

No mortar was needed. Mud and wads of sod plugged the holes. Wagon sideboards formed the roof, overlaid by sod.

People are still living, some with only a few gray hairs, who remember living in sod houses. An aristocracy of sorts--the Sons and Daughters of the Soddies of North America--has developed. To qualify, you must have been born in, or lived in, a soddy. Far from a dying breed, the membership increases as more old soddy residents find out about the S&D of S of NA. Who knows--perhaps having lived in a soddy will become a snob symbol like coming over in the Mayflower!

You can see a real live soddy at Madison's Prairie Village. A Madison school principal, Miss Bena Jacobs, lived in one from March to October, 1910, in Alberta while her family's proper wooden house was a-building. Far from considering her childhood deprived, she says, "I've always been thankful to my parents for giving me this heritage. I want other people to have the experience of seeing how pioneers lived."

So when Prairie Village started in 1966 she volunteered to build a genuine soddy there. It's pretty big and rather ornate, as soddies go, being 20x13½ feet inside and having a pole frame. The walls are 2 feet thick (some old-time soddies were 3 or 3½ feet thick, she recalls) and 7 feet high. "We built it higher than most because we wanted visitors to move around comfortably."

These thick earthen walls are excellent insulators, keeping out summer heat and winter cold. And they're surprisingly durable. "I recently met a woman born in one in Nebraska in 1901 who says it is still in good condition," she says. Barring accidents or the itchings-scratchings of cattle, and if the builder locates the building on high ground where toad-strangler runoff won't lap the wall, the building is good for generations.

How to build a soddy: Miss Bena Jacobs, who lived in a Canadian soddy when she was a child, shows how you build the wall. "I don't ordinarily wear my Sunday clothes for this job," she chuckles.

A fringe benefit of soddy living is the "wall garden" you can raise when prairie lupines, cacti, other wildflowers sprout out of sod wall.

Soddy even has a real storm cellar such as protected settlers from tornadic "cyclones" and kept vegetables and milk cool. Note weeds growing in roof sod. Here you can see how the gable ends are finished--another unusual feature. Most soddies had shed roofs--simpler to build.

Soddy is furnished with a real cast-iron cookstove, just like granma used before the REA came through, and with other authentic pioneer furniture.

The roof, though, gave trouble, always. "I remember two little leghorn straw bonnets my sister and I were quite fond of, brought from the States," recalls Miss Jacobs. "A long hard rain soaked through the sod roof and stained our hats. But mother got out the sadiron and heated it on the topsy stove and cleaned them up and made them pretty again."

Miss Jacobs gives some handy household hints on building a soddy: "Choose virgin prairie sod——buffalo grass is best because the curly blades give extra adhesion." She's never tried domestic sod but scorns it, considering even sodbound bromegrass too friable for construction.

"Plow with a regular breaking plow such as the pioneers used, as deep as the roots are matted." She had to settle for 3 inches at Madison because the over-grazed roots didn't go any deeper. You're supposed to use oxen, but in our benighted space age you may have to settle, as she did, for a horse. The horse, borrowed from under the saddle, disapproved of the entire project.

You don't need a fancy foundation. The good way is to lay your base course directly on the sod and to build from there——eschewing, of course, low or boggy spots. Choose a knoll or slight slope above high water. The plot she was assigned was not firm enough, so she laid the base course on gravel fill in a trench.

Sometimes the builders simply rolled up the sod strips for transport to the building site, then unrolled them on the rising wall. It's important to place the sod upside down on the walls, right side up on the roof, Miss Jacobs says.

Bridge the roof with tree branches, brush, boards or the sideboards off your wagon——whatever comes in handy. Roof can be peaked or shedded, whichever suits your artistic fancy. Overlap the sod strips like shingles to shed rain, starting at the eaves and working toward the peak.

The housekeeper needn't worry about dirt crumbling off the inner walls, once they've settled, Miss Jacobs assures the nervous pioneer mother. For exhibition purposes she finished the walls of her soddy in the several ways of our forebears: clay, plaster, heavy paper and boards, and left as-is. She floored part of the soddy with boards, part with the usual clay, which she says, "is clean as can be."

The Prairie Village soddy is furnished with authentic period cookstove, bunks, churn, kitchen "safe" (cabinet), other genuine pioneer homestead articles. It can be inspected whenever Prairie Village is open—— every day from Memorial Day through Labor Day.

Also at Madison's Prairie Village you can see an original claim shanty. The Homestead Act of 1862 gave 160 acres (¼ square mile, considered a family-size farm at the time) to anyone who would farm it and build a habitable dwelling thereon. Most of the Great Plains were settled that way——people still claimed homesteads into the '20's.

The law was rather lenient in its interpretation of a "habitable dwelling." Most of these were, if not soddies, mere shanties of 6x8 feet or so made of rough boards hauled 100 miles or more from the end of steel. The door was of planks, the floor of dirt, and the single window a sliding wooden shutter. Anyone rich enough for a glass window was the aristocrat of the community.

The claim shanties and often the soddies were heated to a small degree by a cast-iron stove which also did for cooking. The settler had his choice of fuel——prairie hay or buffalo chips, and later the dried dung "platters" cattle so thoughtfully make for us.

The trouble with grass was it blazed up in fury and was gone. Then somebody invented the air-tight stove in which the grass, twisted into a tight bundle, would glow and give off heat for a longer time. Tying hay into the necessary knots was an all-family job to which even the little ones could contribute. Stories were told of families spending the entire winter just tying hay to keep from freezing.

That's how Hayti, S. D. (pronounced HAY-tie) got its name. Settlers gathered in the marshes there to tie the lush marsh grass into fuel bundles.

Many a claim shanty stands yet on the plains. Some are vacant and rotten. Others were added to, room by room, as the family prospered and today form the nucleus of a respectable, modern farm house.

Some of these built-on structures even have air-conditioning, now, and electric heat! A far cry from hay and buffalo chips.

Here's what the sod "bricks" look like up close. Pole frame construction is unusual in a soddy; was built here for extra durability.

A real topsy stove: these air-tights kept tight-knotted prairie hay twists burning a long time instead of just blazing up and away. Rubber band holds hay knot tight for display.

Where your groceries come from

Ask city folks where their food comes from and likely they'll say, "From the store."

And there's the chap who'd rather drink milk from a clean carton than from a dirty cow. And probably someone, somewhere, knows not the chicken is blood kin to the egg.

America was an agricultural nation from the first. Indian corn pulled the Pilgrims through the first winter. Indians showed them how to raise their own, though the colonists had no plow and for 12 years tore out bushes with their bare hands.

America remained an agrarian nation until about 1900, when industrialization overtook tillage. Economists ascribe part of our phenomenal industrial development to our no-less phenomenal agricultural progress which enabled one man to do the work of many, releasing more hands for manufacturing, teaching, research, service.

American agriculture produced surplus food to keep prices down, surplus labor to keep wages down. "A successful farm policy," it has been termed.

Spacemen today eat food grown on Earth. Space crews and lunar and planetary colonies will live on organisms "farmed" in lighted space-borne water tanks. (The meal-in-a-pill, scientists tell us, must ever remain what it is today——science fiction.) As long as men eat, farming will be vital.

Man progressed little while he lived by chasing game and grubbing grubs and roots. It was only when he learned a seed poked into a stick-made hole meant food for winter. Then he had to stay in one place to guard his crop and to prune back the invading jungle. Then villages formed, and towns and cities. Food stored up gave man time to think, to invent and to worship gentler gods.

Archeologists think holes poked in the ground were followed by a pointed stick pulled by wives or slaves, later by beasts of burden, scratching a groove. The crooked stick is still the major farm machine in undeveloped countries.

From the crooked stick developed the wooden plow with a curved wooden moldboard which actually turned a layer of soil over, sort of, burying weeds. And some forgotten genius tied an iron point to it with rawhide.

Centuries later a medieval Einstein reasoned out the iron edge.

From the invention of the wooden plow in antiquity and the yoking of oxen, buffalo and camels to it, agriculture progressed little until the horse collar was invented just a few centuries ago. Heretofore the horse pulled against a chest strap which, in a heavy pull, cut off his wind. Thus, though he could trot with a light buggy, he could not even walk far pulling a heavy plow or cart against his windpipe. The Collar, by distributing the load on his broad, firm shoulders, enabled him to pull the plow and wagon. At that time the latest farm technology was fast-stepping horses to replace slow-plodding oxen. An Englishman wrote a book, "Horse-Hoeing Husbandry," pushing this big new technological breakthrough.

Next time you're in a museum, look over the wooden spades, scoopshovels and pitchforks, plows and planters our forefathers used. One wonders how they got anything done at all.

The colonial plow was a massive thing of wood requiring "...a strong team, a stout man to bear on, another to hold, and a third to drive," wrote Charles L. Flint, Secretary, Massachusetts Board of Agriculture, 1872. He described a plow popular in 18th-century America as requiring "...a strong man to hold it,

Left: It took two men and a team of oxen to scratch a shallow groove in the ground of ancient Egypt.—J. I. Case Co.

Center: Wooden corn planter was technologically advanced when it was built in the early 1800's.—Photo courtesy John J. Menchhofer

Bottom: Wood beam walking plow pulled by one or two oxen, horses or mules was the pioneer's chief tool. Following in the furrow gave farmers the name of clodhoppers.—J. I. Case Co.

and about double that strength of team now required to do the same amount of work."

Then came plows of which the entire moldboard was cast iron, or cast or chilled steel. These met great resistance. "Iron poisons the soil," farmers believed. But some courageous—or foolish—folk tried it and through better tillage as much as doubled their yields. Folk-reasoning did a flip-flop: "Strong iron, strong soil," the village wisemen pronounced.

The iron plow conquered the East and Appalachia and the Piedmont. And it broke the prairie sod.

Trouble came the second year. Once the sod was broken, the iron plow wouldn't scour: moist prairie sod clung to it, causing a drag that exhausted man and beast and kept the furrow slice from turning neat and clean.

Mud on the moldboard stopped our Manifest Destiny westward.

Then a frontier blacksmith in Grand Detour, Ill., John Deere, beat a broken sawblade into a moldboard, 1837. The steel plow scoured in the black prairie soil, spearheading modern agriculture. Our western march continued as a broken and bent sawblade turned the prairie sod so corn and wheat could spring up in its track.

J. I. Case Co. Photos

Cast-iron plow wouldn't scour in prairie soil.

People thought only a lazy man would ride a plow.

Every westbound wagon carried a breaking plow.

Foot lift: great back-saver. Note rolling coulter.

From flail to freedom

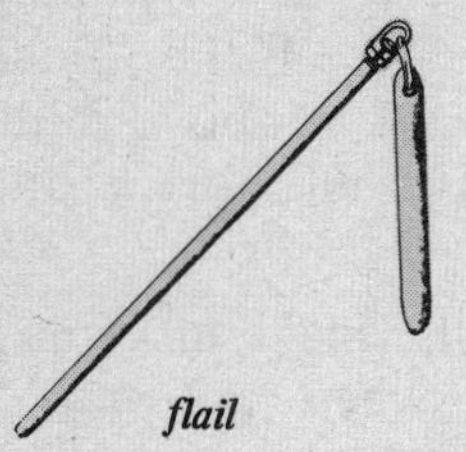

flail

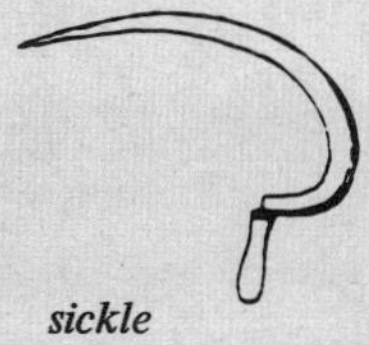

sickle

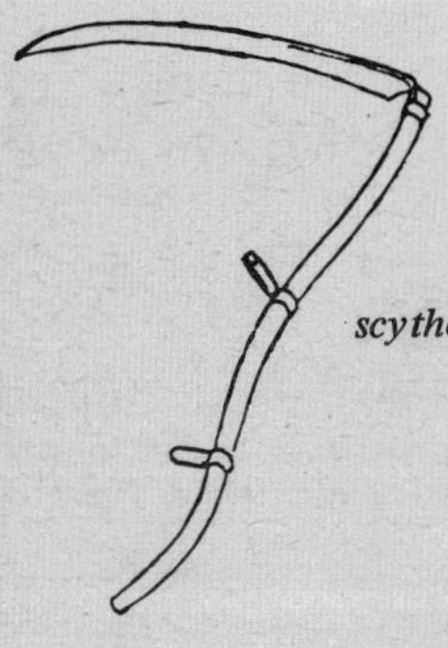

scythe

cradle

As the 1960 USDA Yearbook of Agriculture, "Power to Produce," says, "The progress men have made during the past century in methods of harvesting crops is the progress from flail to freedom."

We know from archeologists that a stone knife first tore heads from stalk. The bronze knife worked better. The Egyptians in 3000 BC used the notched sickle, a small improvement.

They and the Romans threshed by spreading the cut straw on the ground and driving livestock over it, rolling wheels over it or dragging sleds or stone––or iron-knobbed planks across it. Else they pounded it with the flail––a whip-like leather-hinged stick.

A strong man in 10 hours of flailing could knock 7 or 8 bushels out of the heads (and yet not get it all, by far).

The only improvements in the next 4,800 years were the scythe (a "stand-up" sickle), and then the cradle: wooden fingers attached to the scythe which laid the cut grain straws in a row or in bunches for easier garnering. While the men cradled the grain, women gathered it up and, with twists of its own straw, tied the bundles or sheaves.

So incomplete was the harvesting that the village poor survived by gleaning what the harvesters missed.

Thus crudely and incompletely threshed out on the floor, the grain had to be winnowed: the straw raked away and the residue scooped up in baskets and sprinkled out so the wind could carry away the chaff. Our colonists threshed and winnowed just as the ancients.

Plowing and planting and hoeing were work, but a man and his family could take care of enough land to feed themselves through the winter and a smidgen over to sell and trade so city folk could eat, too. It took 13 Revoluntionary farmers to support one city dweller. Today the average farmer produces enough food for some 30-plus people––a figure that will rise 'twixt writing and reading.

With steam power and gang plows and marvelous new binders and threshers, by 1910 a farmer needed only 135 hours of labor to raise 100 bushels of corn, 106 hours for 100 bushels of wheat, and 276 hours for a bale of cotton.

These 1960 averages for the U. S. were about 23, 17 and 77, respectively.

The big bottleneck came in harvest. A man could raise far more grain than he, his family and his neighbors could gather into barns. "...but it (grain) stood so long on the ground before it was reaped it was most shaken...." as the Virginia colony governor reported in 1617.

Groundhog Thresher

Hand-cranked spiked cylinder beat some of the grain out of the heads in mid-1800's.

Then Meickel, a Scotchman, invented the groundhog threshing machine in 1787: a simple toothed cylinder cranked by hand that beat the grain out of the heads. Someone had to rake away the straw, and the grain was basket-winnowed from the chaff as in 3000 BC.

A patent was issued in 1837 on a "combination thresher" which mated the groundhog to a fanning mill. EUREKA! A real threshing machine! Somebody still had to rake away the straw and winnow, but now animal power could be applied: dogs, goats and even horses on treadmills.

"Models made in 1850 could thresh several bushels of wheat an hour with a four-man crew and 6 to 8 horses on the sweep power...." recalls the 1960 Yearbook. "A larger model 10 years later was operated by 8 horses and had a capacity of 300 bushels a day."

The sweep power was a horizontal wheel attached to long poles like wagon tongues which horses pulled around in a circle at 2½ RPM. It was geared to a tumble-rod at 101:1 ratio. Later sweeps harnessed 14 horses.

As the thresher developed in the middle third of the last century, thoughts turned to a better way to cut the grain. Two reaping machines were invented in the early 1830's. Cyrus McCormick's reaper, pulled by one horse, cut the stalks. A man raked them off the platform in bundle-size bunches, as did the cradle, just right for hand-tying. The reaper reduced harvest labor 50 percent and helped insure a more timely harvest so less grain was lost to wind and weather. Later models bunched the stalks mechanically.

Soon after the reaper appeared, a Michigan farmer's wife, it is said, had a dream which impelled her husband to unite the reaper and thresher into one machine: the combined harvester-thresher, now called the combine (pronounced COMbine) which cut and threshed and separated and winnowed all in one operation. Says the Yearbook:

"The machine had all the major features of the present-day combine, but the idea was almost a century ahead of the technological progress necessary to make the machine a success." Even so, for 10 seasons it cut and threshed up to 25 acres a day but its inventors gave up for lack of power and too much moisture in the grain.

Shipped around Cape Horn, the Michigan combine harvested several hundred California acres in 1854 where the fields were bigger and the air dryer, spawning teeming generations of ever-bigger combines. Factory production began in 1880.

Early California combines, weighing up to 15 tons and pulled by 40 horses, cut a 35-foot swath. A ground-driven cleated bull wheel turned the mechanism. As steam developed, in 1890 a steam engine was mounted to turn the works while horses merely pulled the combine. When steam tractors came along, they replaced the horses. Gasoline engines began displacing steam about 1912.

Adding fanning mill and treadmill power to the groundhog boosted production to 300 bu./day in good going by 1855.--J. I. Case Co.

Horse sweep applied power of 6 to 14 horses to the thresher--a great advance.--J. I. Case Co.

The self-binder was first patented in 1850. It not only cut the stalks but tied the bundles with wire, later supplanted by twine, and survived into the 1940's.

Even family-farm-size combines, cutting 10 or 12 feet, required an operator in additon to the tractor driver. The one-man pull-type combine appeared in 1935, cutting a 5 or 6-foot swath, pulled by a 2-plow tractor and powered from its power takeoff, a rotating shaft that transmitted engine power to an external machine. The first machine adapted to the PTO was the binder––in 1919.

Self-propelled grain combines appeared on American farms for keeps in 1938 and steadily gained favor. WW II pressures made their place secure, supplanting pull-types on all but the smaller farms. Self-propelleds are most maneuverable and open a field with less waste, nor do they keep a tractor tied up.

Ideas die if their time has not come. In the late 1920's the Gleaner combine folks built a combine around a Fordson tractor. Farmers laughed, "I've got one tractor. I don't need two. And I sure don't need one I can only use two weeks a year."

Today you can buy self-propelled just about anything: windrowers, hay balers, forage choppers, corn pickers. Some manufacturers make the non-tractor: a three-wheeled power unit which pulls nothing. It forms the power base carrying and powering a combine, etc.

Thus we have seen that muscle ache and the taste of sweat were the farmer's lot––and his wife's––until steam left the factory and the rails to squat, motionless, in the farmer's barnyard to turn the few crude machines available. Putting wheels under it was the next step. A horse-drawn steam engine, the Forty-Niner, was built in 1849 in 4, 10, and 30-HP sizes. The 4-HP jobby weighed 1,000 pounds per HP.

Plowing, the heaviest farm work, got steam attention first. A self-propelled traction steam engine appeared in 1855. In 1859, Abraham Lincoln encouraged steam power on the farm.

Old woodcut shows horse-steered self-propelled steamer and web-stacked, hand-sacked separator. Notice gleaners with wooden rakes working the fields in this late-19th century scene.––J. I. Case Co.

Threshing from the stack near Ramona, S. D., in 1888 with a 14-horse sweep. Separator is hand-fed. To convert this machine to steam drive, the Cheney family took the side gear off and put on a belt pulley. Cheney tried to buy a self-feeder but couldn't get one. A neighbor named Woodmanse took a cardboard pattern off another machine and had the parts cast in Watertown, S. D.--Cheney collection.

Above: The self-binding reaper not only cut the grain and formed the bundles, as did the reaper, but it tied them--first with wire, later with twine--for curing out in the shocks before threshing. Albert and brother Chester Cheney posed on their binders in 1903. Droves of itinerant laborers "followed the harvest" from Texas into Canada. Did the farm term "bundle stiff" father the hobo's appellation "bindle stiff"? Photo is marked HARVESTING THE GOLDEN GRAIN--RAMONA, S. D.--A. B. Cheney collection.

Left: The web stacker was a chain raddle rake that elevated the straw--a big improvement over pitching it away by hand. This 1876 Farquar hand-fed hand-sacked wooden thresher belongs to the Pioneer Club of Indiana. --Photo courtesy John J. Menchhofer.

One man in the radio-equipped air-conditioned cab of a self-propelled combine can "make" as much grain in a day as a whole crew of his fathers.—J. I. Case Co.

Self-propelled steam engines were simply portable power plants built to move themselves about, still steered by horses (that's why you'll see a seat on the smokestack of some early engines). This experience ruined many a good team. The horses expected every load after that to pull itself. All they consented to do was steer.

English tractors had steering gears by 1863 but the first American steering gear was patented in 1880. Eventually the grander steam tractors had power steering.

Then the idea of hauling loads and pulling plows beefed up engines' motive gears and they became land locomotives, or true steam traction engines.

Every year, hundreds of wooden threshing machines burned when an overheated bearing set straw on fire. This danger speeded steel thresher development in 1906.

Stacking bundles to "go through the sweat" prior to threshing.

Unidentified photo. Experts believe engine is a pre-1900 Huber return flue. This design, bringing the smoke and flame back to the firebox end into the stack, was hellishly hot for summer threshing, but warmly appreciated for fall corn shelling and wood sawing. Unjacketed boilers baked a man who had to get out and get under.

Early skid engines.--J. I. Case Co.

"Old No. 1" was built in 1869, now stands in Smithsonian. It was simply a skid engine on wheels moved around by horses. Stack folded back, driver sat on that.--J. I. Case Co.

Nebraska's Steam wagons

By 1868 the 60-HP Standish steam plow and a 4- or 5-man crew could hack up to 5 acres an hour, compared to the acre a day of a man walking behind a 2-horse plow. Two years later a six-bottom gang plow was hailed as a great labor-saver because it required but a single operator, and he could double as fireman.

Village handymen of the '70's started hooking up chains and belts to the wheels of portable steam engines that they might propel themselves. Pity the poor engineer whose chain jumped off a hill (even store-bought jobs did that). He had to jump off, too, to save himself while his engine cartwheeled or rolled or smashed.

Bridges were another hazard: full many a brave engineer went down with his engine when it crashed through a horse-and-buggy bridge. Few engineers appreciated the power of steam; hundreds were killed when they let their boilers boil dry and blow up.

Doubtless the dangers, however unnecessary, added to the engineman's heroic halo and, in the eyes of farmers' sons and farmers' daughters, embellished his steamy day of glory.

By 1900, 5,000 steam traction engines were built annually.

A steam wagon train––no rails––stuck out across the unmarked prairie in July, 1862, from Nebraska City, Neb., for Denver, 600 miles away, hauling 20 passengers. Invented by a Minnesota legislator, Maj. Joseph Renshaw Brown, it was powered by four 12-hp steam engines with 12-inch oscillating cylinders driving wheels 12 feet high. It was steered by little 6-foot wheels attached to the front axle by a ball-and-socket joint. The vertical tube boiler was fed from a water tank integral with the body. It carried water and wood enough for 4 hours.

The "Prairie Motor" ascended a grade of 60 feet to the mile hauling 20 tons in cars behind it. Brown envisioned a vast network of Prairie Motor roads across the Great Plains beating the railroads to the business.

Headed for Denver at a larrupin' 5 MPH, the "beast of the prairies," as some called it, negotiated two steep hills west of town and broke a crank at Mile 12.

His New York foundry having turned to "top secret" Union military work (which turned out to be the "Monitor"), Maj. Brown didn't get back onto his project until after he'd fought the Civil War. By then the railroads had beaten him to the prairie business.

The two front wheels and a cylinder, and a Nebraska City street named Steam Wagon Road, are all that remain of the Major's major opus.

The beast carried wood and water for 32 miles. One question was never answered: where, on the arid, treeless plains, was he going to get more?

Stack folded for low bridges, for storing in shed. This 1885 straw burner was self-propelled and had a steering gear, but could also be horse-drawn.--J. I. Case

The reason you see so many Case steamers around yet today is that Case was the all-time champion steam engine builder, turning out 35,737, nearly three-times runner-up Huber, with 11,568 (figures from Maurice Kildare in RELICS magazine, Winter, 1967). He says Ault-man–Taylor built 5,870; Geiser 15,801; Minneapolis 7,981; Port Huron 6,030, and Harrison 839. The grand total steam engine count when production stopped in 1925, adding up the above, hit 83,824.

However, Kildare neglects to count Russell, Reeves, Nichols & Shepard, Shay and others.

Sawmill woodcut dates from about 1882.––J. I. Case Co.

First the belt

The goat, sheep, dog and later the horse treadmill and sweep accustomed farmers to labor-saving belt power The steam engine took over first as a stationary belt drive. Sawing cordwood and lumber made much use of steamers, which could be fueled free on sawdust, edgings, bark and scraps.

F. R. Howell sawmill built in 1902 has a 48-inch blade. Owned and operated by Milton Ayers, Madison, S. D., shown at the controls (above). Powered below by a 1917 60-HP Case 20-60 at Prairie Village. A shingle splitter is also demonstrated on Show Day.

then . . . . .

Ere the turn of the century, the steam traction engine was well enough developed it could not only power the thresher on the belt but could pull thresher and tender to the next job, and plow between threshing seasons.--J. I. Case Co.

..... wheels under the belt

It didn't really take this many men to water an engine, but this is how popular steamers had become by 1956. This 65-HP engine is pulling a 10-bottom plow.--J. I. Case Co.

A 60-HP Russell here pulls 8 John Deere 14-inch bottoms at Prairie Village. It didn't take this many men and boys to operate the plow, either--one man could do it alone while the engineer and fireman handled the engine. Today a modern farm tractor pulling 4 plows can plow as much in a day with only one man. Plow was built about 1911. Engine is owned and operated by Lloyd Hinker, Woonsocket, S. D.

Then ..... draw bar power

The farmer in a hurry today goes for a 120-HP rig with 4-wheel drive and 4-wheel steer. Manufacturer claims "crawler-type traction and rubber-tired mobility." Crabbing all 4 wheels uphill on a slope counteracts sidehill drifting. Steer with front wheels only, with rear wheels only, turn fronts and rears oppositely for short-radius turns.––J. I. Case Co.

4-Wheel Steering

The agricultural revolution builds steam

Just as the stationary steam engine powered the Industrial Revolution––just as the locomotive and steamboat powered the Transport Revolution––so the steam traction engine and the brood of machines and implements it begat spawned the Agricultural Revolution.

For the first time in history, the man on the land could accomplish more than the limit of muscles. The spectre of famine was banished from an entire nation––a miracle transcending anything the mind had conceived before––because now energy came from other than muscle––from long-stored fossil fuels rather than just from the energy a single summer could store in grain, grass and fodder.

Still gathering speed, the Agricultural Revolution has gone fastest and farthest in the United States thanks to the farmer's readiness to try the new-fangled, his willingness to forget Granpap's ways. Never before in the experience of the human species––and even today only in America and those industrialized nations who have followed our example in mechanizing a free agriculture––has food outrun famine.

Only in the last couple of years have we turned from cursing our "surpluses" to blessing our abundance-and-to-spare.

We think today's marvelous machines are a vast improvement over the steam engine. But look at the improvement the steel plow marked over the iron one; the iron over the wood; the wooden plow over the pointed stick.

Here was tremendous power tamed to man's hand: a single steam operator could guide the power of not 40, but 60, 80, even 250 horses! And as so many kings before it, steam came to kingship through revolution—the Agricultural Revolution. For nearly half a century King Steam reigned supreme in all his smoky glory. Steam was big and powered big machines that inspired big ideas.

Remarkable for sheer size if nothing else was a 1900 Best steam traction engine built for a Stockton, Calif., farming company. Its wood-covered drive wheels were each 15 feet wide and 9 feet high. It weighed 82,000 pounds.

One man plants, covers, packs, fertilizes and pesticizes one trip. The ticket for this one man's pile of iron would have bought four or five steam threshing rigs 60 years ago requiring 100 men to operate them.—J. I. Case Co.

Bonanza boom and bust

While steam was King on the Great Plains, Steam was God in the Red River Valley of the North.

The railroad barons of the 70's and 80's—pushing their rails westward across the fabulously rich Red River Valley at Fargo between Minnesota and Dakota, toward the wide Missouri that slashes Dakota into East River and West River countries—needed something to haul.

Settlers and their chattels would make nice west-bound revenue. And what would be more profitable than hauling their wheat back East?

The railroaders came up with a flamboyant showmanship scheme which for sheer spectacularity and giganticism pales the most costly Hollywood extravaganza to half a spark.

They boomed the bonanzas.

Each bonanza they created was a huge farm of multi-thousand acres, exploiting herds of drifting laborers and the biggest equipment made. Big equipment, of course, meant steam traction engines of a size never before seen off the rails. While they used big steam equipment, "they were not big users of steam. They couldn't afford it," Hiram M. Drache, professor of history at Concordia College, Moorhead, Minn., told the author. Drache wrote "The Day of the Bonanza," from which the following material is borrowed, courtesy Dr. Drache and his publisher, North Dakota Institute for Regional Studies.

Steam traction engines pulling 14-bottom plows were seen. The little 12-ton toys popular among family-type farmers, capable of pulling only 8 plows, merited not even the barons' scorn. Some of the bonanza farms were so huge they even had their own railroads and line elevators.

The goal was profit, of course, but this was but a secondary goal. The first goal was publicity to lure hardscrabble New England farmers to the lush, stoneless Great Plains. Journalists then as now loved bigness and noise and statistics. Papers played up the huge farms—as the bonanza barons knew they would—the huge equipment, the huge crops. Settlers rose to the bait and the railroads fattened on their traffic, which was the whole idea of the bonanza farms in the first place.

Early photographs show 43 binders working in one field; 34 four-horse 8-foot binders in another; a horse barn holding 125 head.

Uncle Sam even issued a 2 cent postage stamp in 1898 showing 18 bonanza gang plows, each pulled by 4 horses or mules, in one lineup.

These were the biggest of the big farms and, by the standards of their day, the most efficient—compounded of the biggest and therefore most efficient steam equipment built; hired hands that need not be fed nor paid during the slack season, but cast to Winter's winds.

The bonanzas failed.

They failed although they bought land as cheap as 16 cents an acre when the family farmer had to pay $1.25. They failed though they avoided taxes for years. They failed although their land value jumped an average of 31 percent a year.

Observes a historian, "The natural rise of land prices was sufficient in most cases to make the bonanzas a financial success. An increase of $1 per acre each year multiplied by several thousand acres certainly gave a good book profit. This favorable speculative factor was a major reason for the financial durability of many of these bonanzas."

Uncle Sam issued an 1898 Two-cent postage stamp based on this Bonanza photo of 18 gang plows in one line up.

In 1879 one of the big land companies sold land, for which it had paid $3.75 an acre just a year earlier, for $10 and up. By 1920 this same land was selling for $90 to $115. Wheat ranged during this period from 75 cents per bushel to as high as $2; the bulk of it sold in the $1.10 to $1.40 range. (Today it is running from $1.57½ to $1.97½.)

In 1880 the Toronto Daily Globe editorialized, "Dakota and Minnesota will rue the day when the bonanza farms were introduced," reports Prof. Drache. He quotes William Dalrymple, son of a bonanza founder, in 1917, "It will be better for the state, for the towns and cities of the state...to have a great many small farms in place of one big farm."

Actually despite the much-shouted efficiency of the huge bonanzas, their profit per acre was so small that the smallest land tax wiped it out. As early as 1883 some of them began to break up. The longest-lived survived into the 1920's.

Drache quotes a 1900 writer, "The impression has been circulated far and wide that in agriculture, as in manufacturing, the big fish are eating up the little ones and that the independent small farm is a thing of the past." But after visiting the Red River Valley he concluded, "The great estates of that region are doomed to disintegration. The great wheat ranch cannot compete with the small diversified farm."

Commented our historian, "Labor on the family plan plus the ability of the family to take advantage of some hidden income which was not possible on the bonanzas eventually proved the superiority of the diversified family farm."

And wrote the Minneapolis Journal in 1893, "The bonanza farmer is not a social animal with his neighbors. He leaves the country early in the fall until spring. The silent and almost deserted farms are left in charge of the foreman; the proprietor spends his winter (and his money—dcj) somewhere else."

At this time seven families totaling 35 people were operating 28 sections (17,920 acres), which could have supported 100 families of 500 people.

Prof. Drache disagrees with my question in Catholic Rural Life magazine (3-67): "But in view of the history of ancient empires which rose on the production of their agricultural families and fell when these families were destroyed; in view of our contemporary observations of the failure of Chinese and Russian communes, in view of world history, which shows the large collectivized farm, whether communal or capitalistic, produces more famine than food; in view of these historically recent bonanza experiments here in our own country, why do we permit the present hell-bent race toward corporate collective farms which have never, ever worked?"

First gas tractor of record lacked carburetion and ignition equipment to make it a commercial success in 1892. Case got into the gas tractor business commercially in 1908 and in 1917 was still saying steam would always be preferred in some places for some farm jobs.––J. I. Case Co.

Sneaky · stinky · noisy · little upstart . . .

As steam grew to supremacy, a slippery little infant was trying to grow up enough to challenge King Steam. A Frenchman tinkered in 1678 with an internal combustion engine fueled by gunpowder. Others, too, experimented with gunpowder as well as turpentine, natural gas, coal gas––even coal dust! Drake's 1859 oil well started the petroleum industry and gave the sputtering babe the nourishment it needed to grow up.

The Otto engine, developed in Germany on the 4-cycle principle we use today, was a factor by 1876. Rudolph Diesel, starting with coal dust, patented his compression-ignition engine in 1892.

Just as the first horseless carriages were built by hanging a gas engine under a de-shafted buggy, so some of the first gas tractors were simply steamers with a gas engine substituted for the steam. In fact, it's hard for the uninitiated to tell an early gas tractor from a prime steamer: they both have smokestacks in front, a big hunk of metal in the middle and a cab at the rear, with little wheels ahead and huge drivers behind.

The first gas tractors were built to LOOK like steam engines––even complete with whistle–– on the theory they wouldn't scare the horses so much.

This is the 1915 Case 20-40 that started Prairie Village. It still runs like a charm. If you can't tell the difference between this and a steamer, don't be discouraged. They fool a lot of people and horses on purpose.

. . . if it would start

The first recorded use of the word "tractor" to replace "gasoline traction engine" is found in an 1890 patent. The 1908 Case catalog presents "steam traction engines," but the 1917 edition sells "steam tractors."

Case built the first recorded gasoline traction engine in 1892. It was not commercially practical, however, needing carburetion and ignition equipment not yet invented. Case offered their first commercial gas tractors in 1895, several sizes of 2-cylinder models.

Practically every major farm machine advance was made by a farmer. In his own shed he cobbled up some crackpot contraption which a manufacturer took over, developed, refined and promoted.

A well-driller and thresherman named John Froelich mounted a single-cylinder gas engine on a steam tractor chassis in 1892, building his own traction gears. His contraption threshed for 50 days and worked at temperatures from +100 to –3. The Froelich fathered the John Deere.

They were all self-starters

The war between gas and steam heated up between 1910 and 1921, when there were 168 tractor companies, of which 160 have died. Steam engine builders refused to load their wares on freight cars carrying gas tractors.

The 1917 Case catalog, while offering both steam and gas tractors, prophesied: "We know that in some localities steam has always been and always will be the most satisfactory power. Time will prove."

The same issue offered a 12-25 gas tractor with two speeds—an innovation: 1.75 MPH for plowing and 2.2 MPH for light work. Said the catalog, "Some light duty tractors we know are so designed as to operate up to 4 miles per hour. This policy we believe is wrong, as it is most extravagant and is unproductive of the best results." Now they'll top 20 MPH on the road, loaded.

Steam engines had been rated at 15-45 HP, 20-60 HP, etc., meaning they could deliver the lower power at the drawbar and the higher power on the belt...they used up 2/3 of their power just to move themselves. Notice how much more efficient were the gas tractors: they needed only half their power for themselves. Today's tractors are rated like 60 HP, period, although every farmer knows they'll struggle harder on plowed ground than on firm.

Case's largest gas tractor in the 1917 book was the 30-60 2-cyl. horizontal 4-cycle with a 10-inch bore and 12-inch stroke, operating at 365 RPM. The magneto was "...protected by a rain-and-dust-proof leather hood. The carburetor uses successfully and economically the various grades of naphtha, distillate, kerosene and gasoline."

The horse came under merciless corporate attack. No less an authority than Thomas Edison pointed out the tractor ate no hay on Sundays.

The same catalog went to bat for steam thus: "Tractor farming is a proven success. A farmer with a reliable tractor can farm at less cost, less time, and at a greater profit to himself than farming with horses. ....It is hardly necessary to comment on tractor plowing. Its success has been demonstrated and proved. Plowing the soil deep enough and at the right time is today reckoned as a factor that increases crop production in many cases as much as 100 percent." The plow that required but a single operator in addition to the engine crew was much extolled.

Horse-powered agriculture peaked in the U S. about 1920 when we had 25 million horses and mules, dropped to a 2-million postwar low. The resurgence of riding has brought them back to near the 3-million mark.

...you started 'em yourself! A. D. McCracken (right) had been cranking the 1920 Minneapolis 35-70 for half an hour by horsing up on the ratchet bar and never got a pop. As he stood gasping with his tongue hanging out to his belt buckle I asked him for "one more for a picture"--and as the shutter clicked, the old ogre took off! Prairie Village has an acre and a half of barrel engines of all sizes. Roger Garlick does the honors with his little jewel (left). Note the steamer-type water glass. And if you think that thing on the windmill looks like a Size 150 overhalls, you're right. They're 15 feet from strap to cuff, and as soon as we can find a feller to fit 'em, Joe will be out of office.

Below: Joe Habeger's 1-cylinder Rumely hitched to a 1906 Case 42-inch steel separator. "The tractor's too small for the thresher, but I just pull it around for effect," says Joe.

Steam threshermen found themselves losing out to more enterprising neighbors who bought gas engines and sold their services on a single competitive advantage to the farmer: "You don't have to haul coal."

Says USDA, "...the rate of technological progress had slowed down (1920). The availability of good new land had dwindled to insignificance. One-fourth of the harvested crop acreage was being used to produce feed for power animals.

"If methods had not been changed, many more horses, more men to work them, and much more land to grow feed for them would be required for today's net agricultural output. The American economy of the 1960's could not be supported by an animal-powered agriculture on our...slowly shrinking land base. National progress on all fronts would have been retarded seriously had not agriculture received new forms of power and sources of energy not restricted by biological limitations."

The first gas tractors were as big and heavy and cumbersome and slow as their steamy-breathed kissin' cousins. This iron-mountain type tractor prevailed from 1910 to 1920, but, says USDA, "it could not accomplish the many tasks necessary to mechanize the farm." It could pull a plow in the field, a wagon on the road, or a machine on the belt. Period.

"Manufacturers began experimenting with light tractors suitable only for cultivation...," says the historian. But the farmer didn't want two tractors--a big one to plow and thresh, a little one to sow and cultivate and mow. Many farmers kept heavy tractors and heavy horses well into the '40's. What they wanted was a farming machine big enough to plow and thresh, small and nimble enough to cultivate and sow. They wanted an all-purpose tractor.

The first successful all-around tractor was the International Harvester 1924 Farmall, which established the tricycle "landing gear" as successful. It plowed and threshed as well as it sowed, etc.

Welcome to Iron Mountain

The Minneapolis Farm Motor 35-70 HP was built in 1920 by the Minneapolis Threshing Machine Co. Its 4-cyl. crossmounted engine had a 7½ inch bore, a 9-inch stroke. The machine weighed 23,500 pounds sopping wet, cost about the same as a steamer of equivalent power. Belt and reverse clutch were built into the pulley; a separate clutch on yon side gave it its one speed of about 2¼ mph. This was a road special.

Belt had to be twisted on some engine-thresher combinations so the thresher would run the right direction and thresh instead of unthreshing. Where the engineer had a choice, as with a steam engine that would run equally well in either direction, he twisted the belt to make it run truer and more proof against the wind. Note the pleasure young and old are taking in Joe's monster. That's him in the pith helmet in the cab, left.

The 450-rpm engine was designed to cool with 60 gal. water and actually had a water-circulating pump. Joe uses drain oil for a coolant—finds it runs hotter so kerosene burns better—and oil doesn't freeze. Observes Prof. Joe: "Sixty gallons is a lot of water to put in every morning by the bucket."

The 'Big 4'

The 1911 Big 4 was built in Minneapolis, as were many gas traction engines of the time. Its drive wheels are 8 feet high and it was sold with an 8-bottom plow just like the one it's pulling. The radiator is a section of a fire-tube boiler, flues and all, almost half as big as that of a small steam engine, and holds 70 gallons of water. Cylinders were cast separately with the heads on: an integral or bonnet engine. You started it by jumping up and down on a torque bar hooked to the PTO shaft at the rear. Owned by Milton Ayers, Madison, S. D.

Agricultural engineers with a prophetic bent predict eventual once-over machines that will prepare the seedbed and plant and fertilize and apply one-shot pesticides in one operation. Following them will come the once-over harvesting machine that picks, cleans and processes, perhaps even canning and freezing and packaging the crop in one pass.

Whether you live in the highest-rise apartment or the plushiest suburb, what happens on the land affects you IF you eat, drink, pay taxes and/or wear clothes. And mostly what happens on the land is agriculture.

Despite one metropolitan daily editor's statement that "agriculture isn't important any more"––despite the fact that less than 8 percent of our people are farmers, despite your paying only 17 percent of your take-home pay to feed your family instead of the 33 percent in the '30's when food was "cheap" (compared to the 50 to 100 percent food costs people in hungry countries), this "unimportant" agriculture today produces 70 percent of our new wealth, 70 percent of our industrial raw materials, employs 4 out of 10 people farming, processing and distributing farm production or supplying farmers with the goods and services they need to produce. Farm exports produce a big chunk of our foreign exchange, holding down our deficit of payments, and keeps a brake on inflation.

Steam power first put farming into the big time––and now steam is being urged to replace gas engines in autos to hold down air pollution!

Full circle, anybody?

Nobody seems to know anything about this Hart-Parr, except that it looked like a steam engine after the fashion of the day and had the steering wheel on the right side.

Hart Parr

1913 Rumely Oil–Pull; 1 cyl. 10x12, 5 plow, 2-speed 1.9 or 2.75 MPH @ 425 RPM. It was oil-cooled to get a higher operating temperature for more efficient kerosene combustion. It weighs 14,000 pounds in its stocking feet, says Joe Habeger, its master.

Rumely

In 1920 the Rumely Oil-Pull had 2 8x10 cylinders and was rated at 20-40 HP, still 5-plow. It's still oil-cooled. Owner Roger Garlick is at the controls, getting 2 and 3.2 MPH out of her @ 450 RPM. Stark naked, the tractor weighs 13,000 lbs. Note the progression here to more complexity (double the number of cylinders), smaller cylinders, higher engine speeds and higher travel speeds and lighter weights, a trend continuing today.

His passenger is having a Big Time blowing the whistle: Garlick added a compressor and air tank just for kicks.

Today's tractors of the same plow-class have 4, 6 or 8 cylinders; engines operate at 1,250 to 4,000 rpm, deliver plowing speeds of 4 and 5 MPH and road speeds of 15, 20 and 30 MPH, and weigh 3,000 to 4,000 pounds.

Nebraska test records show that in 1920 the average wheel-type tractor tested delivered 5 drawbar horsepower hours per gallon of gasoline or distillate. The average wheel tractor tested in 1959 delivered 10 DBHPH (but the gasoline cost 4 or 5 times as much!). Diesel wheel-type tractors went from 10 DBHPH in 1935 to 13.5 in 1959.

An early attempt at "motor cultivation" is this 1919 Allis-Chalmers articulated machine that drives from the front wheels and steers by hinging in the middle. Rear unit carried cultivator, plow, drill, planter, etc. It was rated at 6-12 HP @ 800 RPM with a 4-cyl. vertical LeRoi engine. Speeds were 1.5 and 2.8 MPH.

We think we're so modern today with our electric starter and lights, articulated tractors, front-wheel drives and unit-type implements. The Moline Universal had these in 1919. Rated a 2-plow tractor, it had a 4-cyl. vertical en bloc engine, 3½-inch bore, 5-inch stroke. It delivered 9-18 HP @ 1,800 RPM, weighed 3,300 pounds, went ½ to 3½ MPH.

1927 18-36 Hart Parr 2-cyl. crossmounted 4-plow 2-speed 800 RPM kerosene or distillate. George Demary, Winfred, S. D., owner.

Top: The 4-plow 1925 Huber was rated for 4 plows. Its 4-cyl. engine had 5-1/8x6¼ bore and stroke, deliverd up to 3 MPH @ 1,150 RPM, 2 speed. It weighed 6,230 pounds. This one represents an early conversion from steel rims to rubber tires, probably done in the '30's when cutting down steel wheels and adding rims for tires was about all that paid the village blacksmith's rent. Operator comforts were nil.

Center: Gray 20-40 Canadian special built in 1918 didn't have rear wheels––it drove through a chain to a spiked drum that could be filled with water for more weight and traction. Water ballast, they called it. Today's tractors have cast-iron wheel weights plus water in the tires with calcium chloride as an anti-freeze. The 4-cyl. crossmounted engine gave 2½ MPH @ 600 RPM, and a 20-40 HP rating.

Bottom: But farmers didn't want such small tractors. They wanted them big enough to plow, yet nimble and high enough for cultivating and other work. This 1930 Oliver Hart Parr row crop single front wheel was patterned after the 1924 Farmall row-crop all-around tractor which pattern is still followed today, although more usually with close-set dual front wheels.

Our slave minds

The human race has plodded or ridden behind the horse for so many generations that it is still almost incapable of conceiving any other arrangement. Even though it was easier (because of steering complications) to put power on the rear wheels, we had to put the engine (horse) in front on our tractors (plows), trucks (wagons), cars (buggies) and airplanes (!).

Only a minuscule proportion of these have broken away and powered the front wheels under the engine, or put the engine over the rear driving wheels. Four-wheel-drives (even they've been with us in rare brands for over 60 years) are a facing-up to reality and a divorce from the horse. They're becoming more popular.

Perhaps the rear position of ship and boat propellers relates to the sailor having been pushed by the wind for centuries, rather than pulled.

At least one tractor, Arthur D. Cole's Rein-Drive, tried to handle like a horse.

We are still so wedded to the horse, mentally, that most corn, cotton, soybeans, sorghums, etc., are still planted in horse-wide 36 to 42-inch rows. When I was a kid I thought this was because corn wouldn't grow at any other spacing. Only lately it's been pointed out to me that this is the width of a horse. Our horse-farming forefathers planted the rows far enough apart for a horse to walk between. Tractor makers still largely make tractors and implements to fit the horse row! Only in late years have a few brave souls upped yields with rows twice as close.

The human mind is a slave.

John Deere came out with the first mechanical implement lift in 1928, freeing the farmer from the backbreaking chore of levering his implements out of the ground (although steam-lift plows had trailed the steamers). Harry Ferguson developed his three-point hitch in Ireland that turned implement draft into traction instead of depending on sheer dead weight as had tractors up to then. Teamed with Ford in 1939, his hydraulic control and linkage started a trend universal today.

Caterpillar introduced the first Diesel-powered farm tractor in the U. S. in 1931, the first year a radio-controlled tractor was tested.

Generations of tractor-farmers bumped along on their high-cleated steel wheels until the early 1930's when Firestone put air tires on a tractor. Farmers laughed and laughed until:

The famous race driver, Barney Oldfield, the first man to drive a car over 60 MPH (when he commented to the press, "I tell you, gentlemen, no man can travel faster and live!") was hired. At a farmer field day he plowed the infield with a rubber-tired tractor until a good crowd assembled. Then he unhitched the plow, shifted into a special high gear, and drove that air-tired tractor around the racetrack at 60 MPH!

"It would seem that....Arthur D. Çole has developed a tractor embodying the most desirable features in tractor construction," said MOTOR MECHANICS, June, 1917, "as well as to have combined a wheel tractor and crawler type tractor interchangeable on the one power unit." Technical data on the Cole Rein-Drive is unavailable, but an illustration shows it pulling three plows. Thanks to the Iron-Men Album Magazine.

The power plant of the Rein-Drive tractor is underslung between a pair of wheels which are six feet in diameter. Control is obtained through reins.

Farmers stopped laughing and started buying.

Allis-Chalmers experimented with a fuel cell tractor in 1959: propane gas trickling through fuel cells produced electricity without flame, driving motors in the wheels.

Today AC is showing an air-conditioned cab with radio and closed-circuit TV.

Only now are a few wise heads beginning to realize America's tremendous food production potential, wisely used, can be a more potent force for world peace and the defeat of communism than all the bombs and bullets our industrial might can crank out.

Six-cylinder diesel lifts 4-bottom 2 way plow hydraulically, flops it upside down and goes back in the same furrow.––J. I. Case Co.

The combine wrecked the old-time threshing rings. With the destruction of the old sense of community went neighborliness. "Be independent of your neighbors!" shrieked the ads for one-man family-size combines and balers. Thus the farmer became the rugged individualist he'd always fancied himself.

With rugged individualism came ragged individualism and the fragmentation of agriculture, making it easier than ever——one by one——for interests inimical to family farming to pick the families off. This overvaunted independence today prevents farmers from agreeing on farm

policy, bargaining power or anything else——a socio-economic disintegration as suicidal as a bomb-blast.

Now comes the salvage of classic steam power on the farm, the rebirth of arm-on-shoulder neighborin'. Once again neighbor joins neighbor in frolic and work and mutual helpfulness instead of just glaring at the boob tube and sucking beer.

Maybe there is hope for Rural America yet——thanks to half-century-old engines that recall when steam was glory.